Blackwell
Modular
Science

Key Stage 

Metals

Alan Basnett
and Bob Lee

The Blackwell
Science
Programme
5-16

Contents

Preface

We have written this book because metals, with their unique properties, continue to fascinate us. Metals have a bright future and we should all know more about them.

We offer this topic book on metals to teachers whose pupils are studying the common core and optional studies associated with Modular (balanced) Science at GCSE. The book also covers the criteria outlined in various Attainment Targets of the National Curriculum. The book should also interest pupils studying metals without an exam in mind. There are many activities of a practical and data response nature which are designed to harness pupil enthusiasm. Economic, social, environmental and technological aspects of metals are evident throughout the text. The searching structured questions should offer invaluable practice in this vital type of GCSE question.

ALAN BASNETT
BOB LEE

Unit 1
The importance of metals

Where would we be without them?

Figure 1.1 What's missing? What would our lives be like without metals?

Figure 1.1 shows a typical street scene in a modern town except that some things have been removed. All the things made of metal have been taken out from the picture.

Now compare Figure 1.1 with Figure 1.2, which is complete. Try to identify as many items as possible which are made of metal.

Figure 1.2 How many objects can you see that are made out of metals?

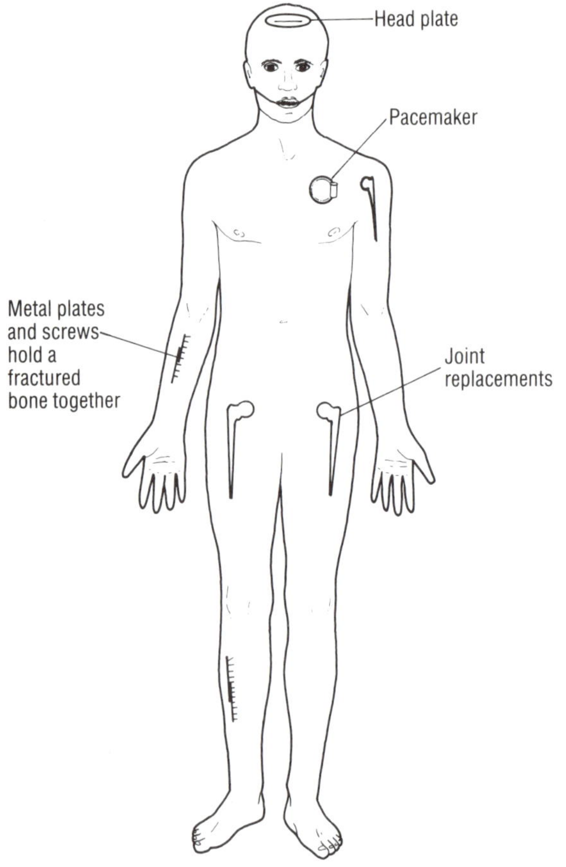

Figure 1.3 Metals in the service of man

Figure 1.4 Some different uses of metals

Figures 1.1 and 1.2 give you some idea of the uses we make of metals. But have you ever stopped to think about just what your life would be like without them?

Start by thinking about electrical things. Electric cookers, record players, TVs, kettles, lights and fridges are all supplied with electricity through wires made of metal. Many parts of the items themselves are made of metal.

The water tanks and pipes in your home are made of metal, too. In the kitchen, the pans, and knives and forks you use every day are made of metal. The clothes you are wearing will have been sewn together using metal needles, probably on machines made mostly of metal. The furniture in your room will be held together with nails, screws and bolts made of metal.

Every form of transport you can think of, from bicycles to supersonic aircraft, is made largely of metal. It is difficult to think of any aspect of our daily lives in which metals are not involved in one way or another.

Figures 1.3 and 1.4 show some further uses of metals.

Why are metals so useful? To find the answer to this question, we need to look at some items made from metal, and then try to think why each one was made from that metal.

We can start with some of the things in Figure 1.2:

- The steel used for making the bridge has to be strong.
- The steel used to make the car bodies has to be capable of being shaped.
- The metal used for making electric cables (usually copper or aluminium) has to conduct electricity and be capable of being drawn into wire.

If we consider some other metal objects, we find that metals have several other properties, too.

- The bell to start and finish lessons is made of metal – most metals are sonorous (they make a ringing noise when they are struck).
- Many people wear jewellery made of metals such as gold, silver and platinum. These metals are used because they can be made into attractive shapes, and because they are shiny – another attractive property. These metals are also quite rare, which makes them expensive, and they also have a special property – they do not corrode.
- Items such as cooking pans and central heating radiators need to allow heat to pass through them, so they are usually made of metals because metals conduct heat.

Uses for a particular metal can change too

The uses for lead in the western world have changed noticeably between 1960 and 1980. It is also possible to predict the uses for lead in the year 2000. Uses of lead include:

- The plates for lead/acid batteries used to power vehicles.
- To surround electrical cables (cable-sheathing) to protect the wires from the atmosphere and in particular moisture.
- As an important source of colours (pigments) for paints.
- Melted with other metals to make alloys.
- As lead tetraethyl, the petrol additive, to give an even burn in the engine.
- Pipes and sheeting for the building industry because it does not corrode and is very dense.

Activity 1.1

To compare the thermal conductivity of metals

For this experiment, you will need a piece of heat-sensitive computer printing paper, measuring about 15 cm by 11 cm. The type which prints in blue gives an attractive result.

1 Collect rods of four different metals, for example aluminium, copper, brass and steel, each about 30 cm long and 3 mm in diameter.
2 Sellotape the rods to the duller side of the paper, as shown in Figure 1.5(a).

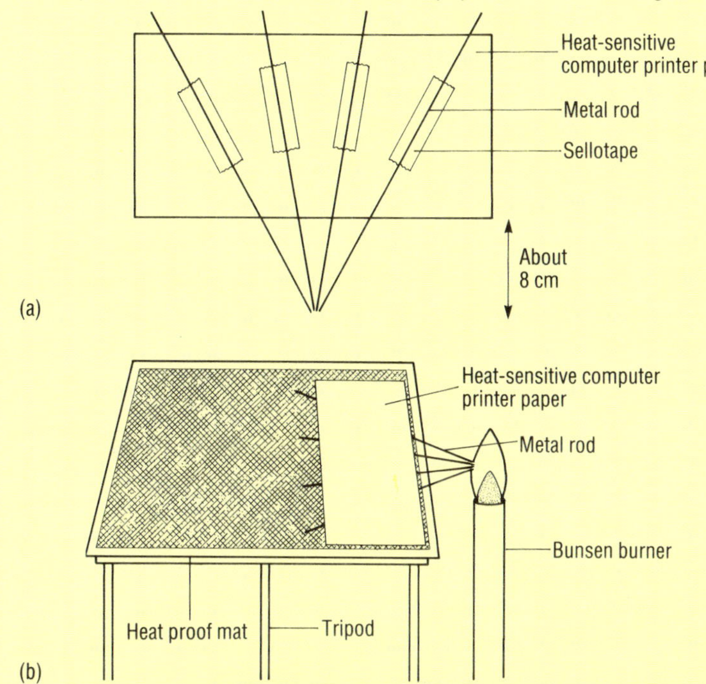

Figure 1.5 Comparing the thermal conductivity of different metals

3 Place the paper and rods on a heat-proof mat on a tripod, with the paper uppermost. Make sure that the ends of the rods, which are close together, stick out over the edge of the mat. Heat the ends of the rods evenly with a Bunsen burner, as shown in Figure 1.5(b).
4 As heat travels along each rod, colour will develop in the paper touching it. When the colour by the best conductor reaches the end of the paper, stop heating.
5 When the rods are cool, remove them from the paper.
6 Label each coloured area with the name of the corresponding metal.
7 Write down the metals in order of thermal conductivity, with the best conductor first.
Stick the paper in your notes.

Activity 1.2

The table below shows the percentage use of all the lead used in the western world for certain years.
Study the table and answer the questions below:

Use	Percentage of all lead used in year		
	1960	1980	2000
Batteries	28	51	64
Cable sheathing	19	8	2
Pigments	10	14	13
Alloys	12	5	3
Lead tetraethyl	8	6	4
Pipe/sheet	14	8	7
Others	9	8	7

1 Draw a pie-chart for the percentage uses of lead in 1960 and in 2000.
2 Which use has shown the biggest percentage increase over the years 1960 to 2000? Give possible reasons.
3 Which use has shown the biggest percentage decrease over the years? What has replaced lead for this use? In what way is the replacement superior to lead?
4 Why is the lead tetraethyl predicted to drop to 4% by the year 2000?
5 A friend tells you that the use headed 'Others' includes making a school bell. Give reasons for thinking the friend is wrong.
6 It is predicted that more lead will be used in the year 2000 than ever before. What will most of this extra lead be used for?

Activity 1.3

To show that metals conduct electricity

You will need: 12 V car bulb in socket, switch, 12 V DC supply, 2 crocodile clips, connecting wires and samples of various types of metals, also wood, glass, plastic and graphite.

1 Connect the circuit as shown in Figure 1.6 using a 12 V DC supply and a 12 V car bulb in its socket.
2 Make sure the switch is off and connect samples of different metals (as wire or sheet) in turn.
3 Turn the switch on. The metal is a conductor of electricity if the bulb lights up.

Do all the metals cause the bulb to light up?
Repeat the experiment with wood, plastic, glass and other materials. Do any of these materials conduct electricity?

Now try with a rod of graphite in the circuit. What is surprising about this result?

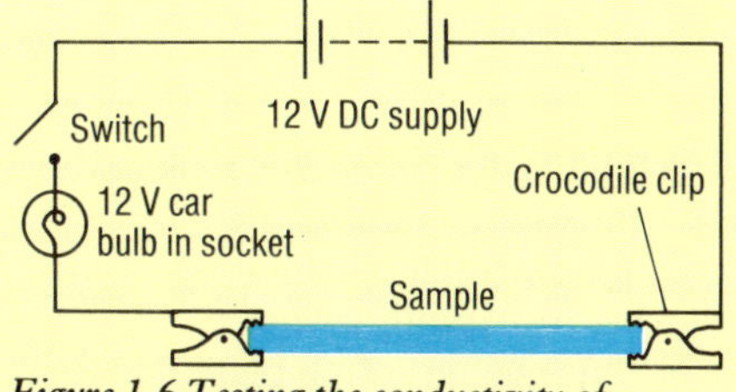

Figure 1.6 Testing the conductivity of different materials

Activity 1.4

Make a table using the properties of metals listed in the Summary horizontally and the use of a particular metal vertically:

Use	Strong	Malleable	Ductile	Sonorous	Shiny	Dense	Conductors of heat/elec	High melting point
Copper for electrical wire			✓				✓	
Copper for saucepans	✓	✓			✓		✓	✓

Look for items made of metal at school, on the way home, and at home. Make a note of each one, and try to find out what metal each is made from.

For each one, fill in the table, ticking the particular property or properties of the metal that make it good for the use. For example, copper is used for electrical wires because it is ductile, and conducts electricity well ... these two properties must therefore be ticked.

Summary

Metals are generally:

- strong
- malleable (can be beaten into shape)
- ductile (can be drawn into wire)
- good conductors of electricity
- sonorous (ring like a bell)
- shiny
- good conductors of heat
- usually dense (they feel heavy)
- solids at room temperature (they have high melting and boiling points)

Questions

1 Mercury is an unusual metal at room temperature. Which of the usual metallic properties does it possess? Can you find some uses for mercury? Say why it is chosen for each use.

2 Try to name the following metals from their descriptions:

a This metal burns very brightly in air and was used in photography before 'flash-bulbs' were invented.

b This metal has a red/brown appearance and is sometimes used for saucepans.

c This metal has a grey look and has a low density. It is used in the fuselage of aeroplanes.

d This metal has been used for fishing 'weights'. It is now being replaced because it kills swans.

e This metal is often plated on to iron/steel parts of bicycles and cars to protect them from corrosion and give a mirror-like appearance.

f This metal is present in some street lamps to give a harvest-yellow light.

Unit 2
Choosing the right metal

Mending a wheelbarrow

Imagine your father's wheelbarrow has rusted through. He looks at the cost of buying a new one, and decides it would be too expensive. He inspects the rusty one, and finds that the wheel, the frame and the handles are in good order. He reckons he should be able to buy a sheet of metal, and make a new body for the wheelbarrow at a fraction of the cost of a new one. He asks for your advice on which metal it should be. How will you decide?

You will need to think about how the wheelbarrow should look, and how it must behave when it is finished, what will have to be done to the metal in order to make it the right shape, and how much it will cost. In other words, you need to make a list of the properties the chosen metal must have. Make your list, and compare it with father's list below.

Father's list:

The metal chosen must

- not go rusty!
- be light but strong
- be malleable (I will have to bend it to the right shape)
- not be too hard (I will have to cut it and drill holes in it to bolt it to the frame)
- not be too expensive
- be fairly easily obtainable

How does your list compare with father's? Do you disagree with any of his points? Do you have any points that he missed? In the rest of this unit, the properties of several metals will be considered, and related to some of the jobs the metals perform. The information should help you to decide which metal you will advise your father to use. But remember, in such situations, there is rarely one 'correct' answer. There are often several possibilities.

Testing metals

To help them decide which metal to use for a particular job, engineers have to carry out a number of tests and measurements (Figure 2.1).

The first tests which are carried out allow the engineer to calculate the forces each part of the design will have to withstand. Then a metal with the desired properties is selected for each component.

The selection may be made from metals already known, or, if none is suitable, a new alloy may be developed having the required properties. Once the new alloy has been produced, it is then tested to check that it is suitable.

Engineers will want to know about several mechanical properties of the metal. They will need to know if the metal:

- resists breaking when large forces are applied to pull a rod of the material (if so, it is said to have high **tensile strength**).

Figure 2.1 Metal undergoing tests on a universal testing machine

- remains unmarked when a hardened steel ball is pushed heavily onto its surface (if so, it is said to have **hardness**).
- pulls into a wire easily (if so, it is said to be **ductile**).
- returns to its usual length after the removal of forces which have increased its length a great deal (if so, it is said to have a high **elastic limit**).
- supports a large load without breaking while showing little increase in length (if so, it is said to be **strong**).

Tensile strength

To measure the tensile strength, a bar of the metal is stretched in a special machine while a measurable and increasing load is applied. At the same time the increase in length (extension) of the piece of metal is recorded. Measurements are taken until the bar breaks. A graph of force (in newtons) against extension (in centimetres) is then plotted.

If your school has a tensile test machine, you can use it to obtain results from which you can plot a graph of force against extension for various metals. A typical force-extension curve looks like that shown in Figure 2.2.

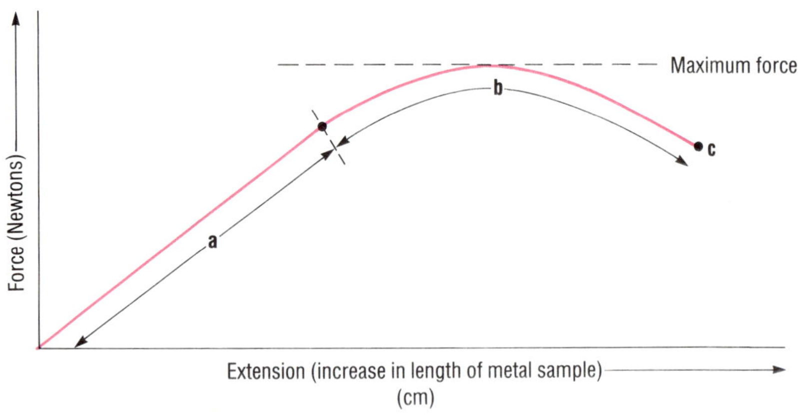

Figure 2.2 A typical force-extension curve for a metal

In region **a**, the metal bar gradually increases in length as the load is increased on it. In this region the metal will return to its original length when the load is removed; the metal is said to be behaving **elastically**. At the end of region **a** is the **elastic limit**. The metal will not return to its original length beyond this point.

In region **b**, the metal extends more easily for a given force and cannot return to its original length; it is said to be behaving **plastically**. The specimen begins to break in this region forming a 'neck' (an area with a reduced cross-section – see Figure 2.3). The extension then takes place only in the 'neck' region. The specimen breaks at point **c** on the graph.

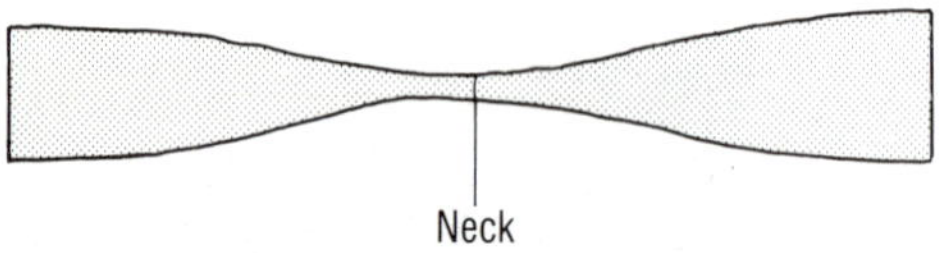

Neck

Figure 2.3 A metal is extended to the point where it forms a neck

Hardness

There are many situations in which it is important to know the hardness of a metal. The metals in cutting tools such as drills, knives, scissors, lathe tools, files, etc, obviously need to be much harder than the material which is being cut.

Activity 2.1

To compare the hardness of different metals

Your teacher will supply you with samples of different metals.

1 Use the end of one sample of metal to try to scratch each of the other metals in turn.

2 Record your results in a table like the one below, putting a tick if the metal causes a scratch, and a cross if it does not.

3 Repeat, using the remaining metals.

Scratching metal / Metal being scratched	Silver steel	Mild steel	Copper	Aluminium
Silver steel				
Mild steel				
Copper				
Aluminium				

Your teacher may supply you with more samples of metals than are shown in the table. Add them to the table.

The column with most ticks will be the hardest metal. The one with fewest ticks will be the softest.

List the metals in order of hardness.

Activity 2.2

To compare the ductility of metals

Your teacher will supply you with samples of several metals. Each sample will have the same dimensions.

1 Clamp the first sample in a vice, and use the lever to bend the sample through 90° (see Figure 2.4).

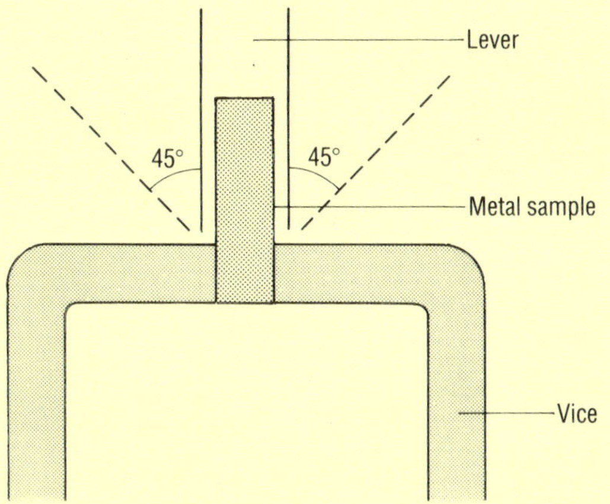

Figure 2.4 Apparatus to compare the ductility of different metals

2 Note the number of bends required to fracture the sample.
The fewer bends required the less ductile the metal.

3 Repeat with the other samples.

4 List the metals in order of **increasing** ductility.

In industry several methods are used for testing hardness (one of which is shown in Figure 2.5). In the Brinell test, a hardened steel ball is pressed into the metal for 10–15 seconds under a specified load. The diameter of the impression made is measured using a vernier microscope.

$$\text{Brinell hardness number} = \frac{\text{load applied to steel ball (kg)}}{\text{spherical area of impression (mm}^2)}$$

The harder the metal, the smaller will be the impression made, and the higher will be the Brinell hardness number.

Figure 2.5 An industrial hardness testing machine

Other properties

An engineer also needs to know about other properties of metals, such as melting point and density. The table in Activity 2.3 shows some important properties of several metals.

Activity 2.3

Study the table below showing the properties of some metals and answer the questions.

Metal	Melting point °C	Density g cm^{-3}	Hardness number (Brinell Number)	Tensile strength 10^7 pascal	Price £ per tonne
Aluminium	660	2.7	20 to 27	5 to 11.4	900
Copper	1084	8.93	45 to 100	22 to 43	1000
Lead	327	11.34	Very low	1.5	350
Nickel	1455	8.91	90 to 210	34 to 99	3600
Zinc	420	7.14	50	13.9	600
Tungsten	3406	19.35	225	12	

The range in values for hardness and tensile strength for a particular metal occur because metals can be treated (eg heating and slow cooling, etc) in different ways and this changes their properties.

1 Which metal has the highest melting point? Is it highest in any other properties? If so, name the properties.
2 I require a metal with a density above 15 g cm^{-3} for a task. Which metal in the list should I choose? How does this metal compare with other metals on properties apart from density?
3 I require a soft metal which melts above 400° C and which is reasonably cheap and strong and not too low in density. Which metal from the list should I choose? How does this metal's hardness compare with the other metals?
4 I require a metal that varies little in hardness and tensile strength when treated. A low density is also most important. Which metal should I choose? How does this metal compare with the others on hardness and cost?
5 The metals vary considerably in price. What sort of factors do you think control the cost per tonne of a given metal?

Some uses of pure metals

Aluminium is made into door and window frames, cooking pans, cooking foil, electric wires and cables

BECAUSE

it is fairly cheap
it has a low density
it conducts heat and electricity well
it is resistant to corrosion
it can be rolled into thin sheets

Copper is made into wires, water pipes

BECAUSE

it is a good conductor of heat and electricity
it is ductile and malleable
it is resistant to corrosion

Tin is used to coat cans made of steel for peas, beans etc

BECAUSE

it is resistant to corrosion
it is malleable
it is non-toxic

Lead is used for plates in car batteries, flashing on roofs, protecting undersea cables, enclosing radioactive isotopes

BECAUSE

it absorbs atomic radiation
it resists corrosion
it has a high density
it is malleable

Zinc is used to coat steel to prevent it from rusting

BECAUSE

it has a fairly low melting point
it is more reactive than iron
so it corrodes in preference to iron

Summary

When choosing a metal for a particular purpose the properties needed must be studied.
Results of tests on

- tensile strength
- hardness
- ductility

help to decide the most suitable metal.
Other factors like the melting point, density, price, etc also have to be considered.

Questions

1 Using information from this unit, you should now be able to advise father which metal he should use to make a new body for his wheelbarrow. Choose from this list: copper, tin, zinc, iron, aluminium, lead. Give reasons for your choice.

2 As a metallurgist, you have been asked to advise in the production of a new drill for drilling through some particularly hard rock. Study the table of properties given in Activity 2.3 and decide which metal you would use for the tip of the drill, the main part of which will be made from hard steel. Give two reasons for your choice.

3 Aluminium and copper are both good conductors of electricity, and both are ductile. In the past, copper was commonly used for making electric wires and cables. Now it is gradually being replaced by aluminium. Suggest two reasons why this should be so.

Unit 3
Mixtures of metals – Alloys

What is an alloy?

Although metallic elements find many uses in the pure state, most metals in use today are **alloys** – mixtures of metals. Alloying alters the properties of metals, and for many jobs alloys have been tailor-made. The proportions of ingredients in the mixture have been chosen in such a way that the properties of the resulting alloy are exactly those required for the particular job.

Alloys of copper

Copper combines with many other elements to make useful alloys. The type of element added dictates the properties of the alloy such as:
strength, colour, resistance to corrosion and wear, ability to machine it into shapes.

Activity 3.1

The chart below shows how the elements can alter the properties. Use the key to symbols for elements to answer the questions below.
Zn=zinc, Al=aluminium, Sn=tin, Mn=manganese, Te=tellurium, Ni=nickel, Fe=iron, S=sulphur, P=phosphorus, Si=silicon, Cd=cadmium, Cr=chromium, As=arsenic, Pb=lead, Zr=zirconium, Ag=silver, Cu=copper.
In each case choose the maximum number of elements that can, when added to copper in an alloy:

1. Increase corrosion resistance but not particularly the strength.
2. Combine increased corrosion resistance and strength.
3. Combine increased strength, and resistance to corrosion and wear.
4. Combine colour, strength and corrosion resistance.
5. Combine machinability, colour and strength.

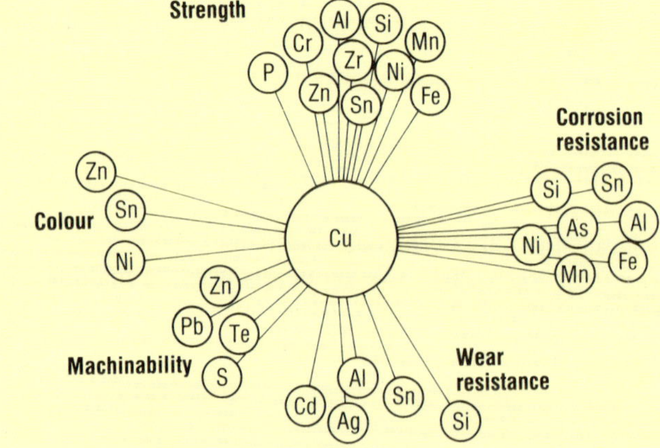

Figure 3.1 Properties of many copper alloys

Activity 3.2

Making an alloy-solder

Your teacher will demonstrate this to you.
THE EXPERIMENT MUST BE DONE IN A FUME CUPBOARD. REMEMBER: MOLTEN METALS ARE DANGEROUS.

(a)

1 Put damp sand into a metal tray to a depth of about 2 cm. Using a wooden dowel, make three moulds in the sand, each about 1 cm deep. Mark them T,L,A.

(b)

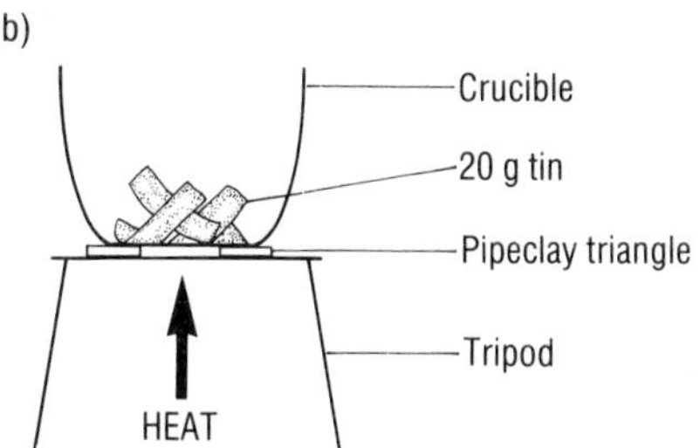

2 Weigh 20 g of tin and place it in a crucible. Heat the crucible until the tin melts.

(c)

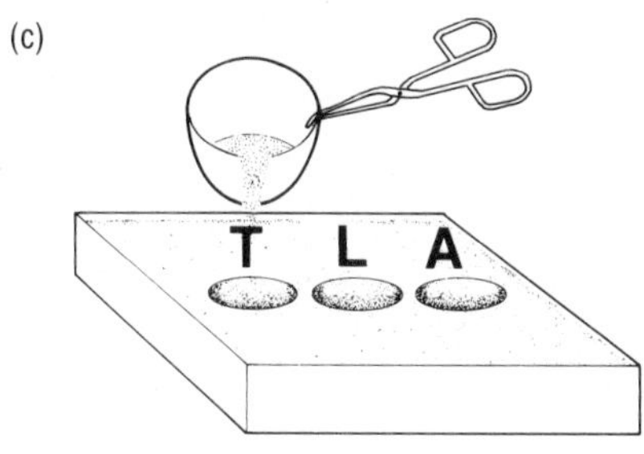

3 Using tongs, carefully pour the molten tin into mould T.

4 Repeat steps 2 and 3 using 20 g of lead. Pour into mould L. (Lead vapour is *poisonous*.)

5 Repeat steps 2 and 3 using a mixture of 10 g of tin and 10 g of lead. Pour into mould A. Leave the castings to cool.

(d)

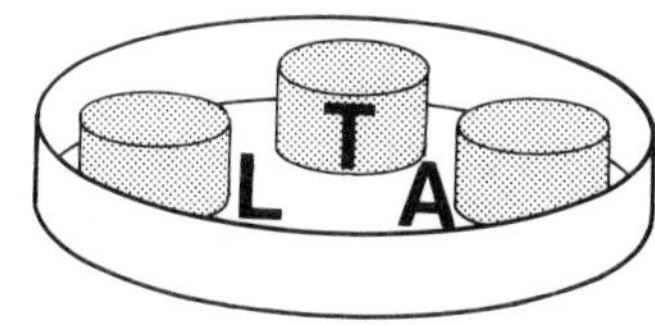

6 Label a tin lid T,L,A. Place the tin casting by the T, the lead casting by the L and the alloy casting by the A.

(e)

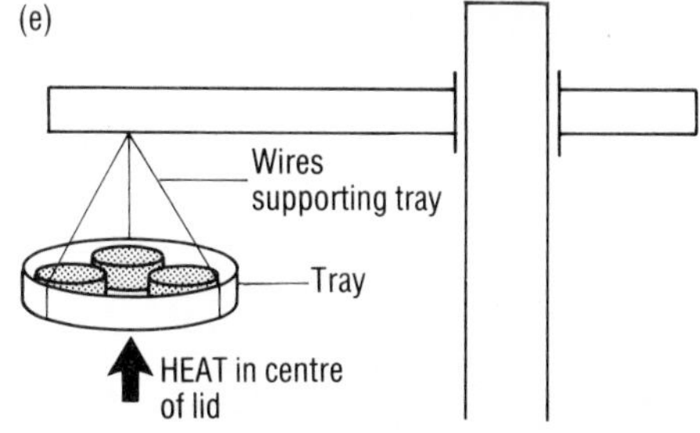

7 Use a heavy-gauge wire to suspend the tin lid from a clamp stand. Heat with a Bunsen burner, making sure the burner is exactly under the centre of the lid. Note the order in which the castings melt.

The alloy your teacher has made is solder.
Answer the following questions.

1 What was the order in which the metals melted?
2 Has alloying affected any of the properties of the tin and lead?

Steel

By far the most widely-used alloy is steel. It is unusual because it is a mixture of a metal, iron, and a non-metal, carbon.

Pure iron is a soft metal. Cast iron from the blast furnace is strong but brittle. (See Unit 9 for a detailed description of the blast furnace.) Cast iron contains about 4% carbon. Items made from steel, such as car bodies, bridges and washing machines, need the strength of cast iron, but not its brittleness. By reducing the carbon content of cast iron, the brittleness is removed, but strength is retained.

The table below shows how the tensile strength of steel increases with increasing amounts of carbon.

Carbon %	**Tensile strength** 10^8 Pa	
0.10	3.7	
0.22	4.8	Mild steel
0.39	5.4	
0.54	6.9	
0.81	7.9	Hard steel
1.04	8.4	

Mild steel is used for making car bodies, washing machines, etc. Hard steel is used for tools such as hammers and drills.

Figure 3.2 A duplex brass fitting showing the soft, white 'meringue' corrosion caused by zinc carbonate

Figure 3.3 Conventional brass (left) shows considerable dezincification after testing. The sample of dezincification-resistant brass on the right shows no evidence of attack after exposure to the dezincification test

Brass

Look at the tubes and fittings that distribute water in your school and at home. The tubes will usually be made of copper. The fittings such as joints, bends, stop-cocks and taps are made as castings. A form of brass known as duplex brass is commonly used for these castings. Duplex brass contains 60% copper and 40% zinc by weight.

In a few areas of the country the duplex brass fittings were leaking and sometimes fracturing completely. A closer look at the failing parts showed growth of a very light, white powder (Figure 3.2). It was not the usual limescale formed in hard water areas. It proved to be a substance called basic zinc carbonate.

Further detective work showed that a particular combination of chemicals in the water was causing the problem. These chemicals were harmless but for their effect on duplex brass. The areas affected generally had soft water which contained some temporary hardness from limestone. They also had some chloride in the water. This combination of ingredients was leaching out zinc from the duplex brass fittings (known as dezincification!). The zinc became the light, white basic zinc carbonate. The duplex brass gradually corroded resulting in a blockage, leaking and sometimes fracturing.

The answer to the problem was found to be a slightly different alloy of metals known as dezincification-resistant brass (Figure 3.3). The composition of this alloy is 61.4% copper, 36% zinc, 2.5% lead and 0.1% arsenic by weight. This brass is also heat treated in a special way. Most of the water fittings in areas with this water problem are made from this special brass.

Questions

1 What happens to duplex brass in dezincification?
2 What must the water contain to cause dezincification?
3 What is different about the composition of the new brass that overcomes this problem?
4 Why do you think water pipes are made from copper but the fittings are not?

Alloys make money too

In Roman times coins were made from lead because it was easily cast and resisted corrosion. Silver and gold were also used for coinage. Silver remained as part of British coinage until 1947, combined in an alloy with copper, nickel and zinc. Then the silver was removed and extra copper was added to give the coins more strength. All of these metals were rather soft and easily scratched.

Bronze coins, containing tin and zinc with copper, followed as a direct result of the French Revolution. The revolutionary atheists destroyed churches and tried to find a market for bronze church bells. An equal weight of copper was added to the bronze. Then both metals were melted together. This alloy makes a good coinage metal because it is hard and shiny and is not easily tarnished. During the years more copper was added until the mixture became 95% copper, 4% tin and 1% zinc. Britain adopted this mixture from France in 1860. Our modern day 2p and 1p coins are similar to the recipe adopted in 1860.

How are coins made?

The alloy is melted in an electric furnace. Then the molten alloy is poured into a casting machine and emerges as slabs about 13 mm thick. The slabs are rolled into strips; sometimes a softening process is involved at this stage. Discs are punched out and softened on a belt passing through a furnace. The discs are cleaned and 'struck' between dyes to give the pattern and words.

Some other alloys

Alloy	Made from	Special properties	Uses
Brass	70% copper 30% zinc	Hard, but easily worked Does not corrode	Musical instruments Electrical contacts Ships' propellers
Bronze	90% copper 10% tin	Harder than brass Does not corrode Sonorous	Statues, ornaments, bells
Duralumin	95% aluminium 3% copper 1% manganese 1% magnesium	Light but strong	Aircraft parts
Nickel steel	95% iron 5% nickel	Hard and tough	Guns and axles
Stainless steel	74% iron 18% chromium 8% nickel	Does not rust	Car parts (springs, electrical casings) Kitchen sinks Cutlery

These proportions have been worked out experimentally, by 'trial and error'. It is very difficult to predict in advance just what properties a particular alloy will have. Proportions are varied, and the alloy is tested until the one having the desired properties is found. In many cases, the alloy chosen may not be absolutely perfect, but may offer the best compromise of properties.

Activity 3.3

Study the table and then answer the questions below.

Coin	Copper	Nickel	Zinc	Tin
£1	70	5.5	24.5	
50p (also American 'nickel' 5 cent)	75	25		
20p	84	16		
2p and 1p (same as old pennies)	97		2.5	0.5

1 How does the composition of the modern 2p and 1p coins compare with that adopted from France in 1860?
2 Why do you think that different alloys are used for different coins?
3 Is the name 'nickel' suitable for the American 5 cent piece? Suggest a better name for the coin.
4 Which element is contained in all of our modern coins?
5 Which element gives our modern 'silver' coins their distinctive look? When was silver phased out of British coinage?
6 What properties must an alloy have to make coins with a long life?

Summary

An alloy is a mixture of different metals melted together.
The addition of one metal to another to make an alloy can alter

- strength
- colour
- resistance to corrosion
- resistance to wear
- machinability

Brass is an alloy of copper and zinc.
Bronze is an alloy of copper and tin.
Solder is an alloy of lead and tin.
Steel is an alloy of iron with another metal (eg chromium) and carbon.
Our coinage is made from alloys including two or more of the following metals: copper, nickel, zinc and tin.

Figure 3.4 Much of this aircraft is made from an alloy called duralumin

Questions

1 Using the figures given on page 12 plot a graph of tensile strength against % carbon for steel.
What effect does the increasing carbon content have on the tensile strength of steel? What carbon content would give a tensile strength of 6×10^8 Pa?

2 Much of this aircraft is made from an alloy called duralumin.
a What is meant by an alloy?
b Name the metals used to make duralumin.
c Why is duralumin used in preference to pure aluminium?

3 Find out what silver steel is made from. Why is it called silver steel?

4 Find the eighteen different metals and alloys in this wordsearch. The words may be read vertically, horizontally or diagonally. The solution is on page 62.

S	S	M	U	N	I	T	A	L	P	N
M	T	S	Z	I	N	C	T	E	N	S
U	C	E	A	T	G	O	L	D	I	Y
I	H	S	E	R	S	P	E	L	L	R
N	R	O	D	L	B	P	V	L	I	U
I	O	D	A	E	I	E	H	E	T	C
M	M	I	Y	A	R	R	T	K	H	R
U	I	U	G	D	O	Z	O	C	I	E
L	U	M	A	G	N	E	S	I	U	M
A	M	U	I	C	L	A	C	N	M	S

Hidden words are
PLATINUM
STEEL
ALUMINIUM
BRASS
SILVER
LEAD
GOLD
MERCURY
MAGNESIUM
LITHIUM
IRON
TIN
COPPER
ZINC
NICKEL
CHROMIUM
SODIUM
CALCIUM

Unit 4
How metals are shaped

Several methods are used industrially to make metals into different shapes. This unit describes some of the different ways in which metals can be made into objects of any shape or size.

Sand casting

The diagrams in Figure 4.1 show how a small metal object can be made using a mould formed from sand. **Sand casting** is used to make some quite large things like the propellers for ships. A mould of silica sand and portland cement is made. Because they are so large, pits are used for the moulding, for safety and to bring the top of the mould to a reasonable height for working.

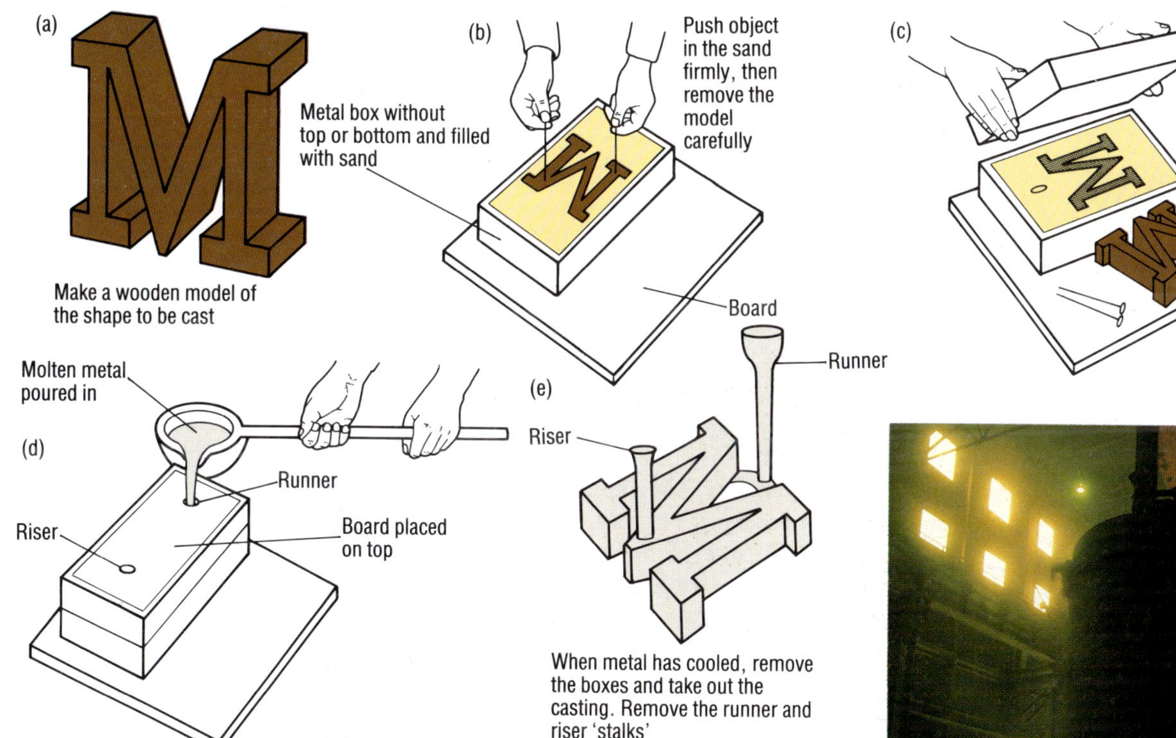

Figure 4.1 Sand casting a metal object

Die casting

Making a mould from sand each time a casting is required is very time consuming and expensive. Castings involving aluminium, copper, zinc or their alloys are often made in **metal dies**. A metal die consists of the pattern. The molten metal is poured in (Figure 4.2) and the two halves of the die are closed. When the metal is cool the halves of the die open and the casting can be removed. The die can be used many times.

Figure 4.2 Pouring steel into ingot moulds

Other ways of shaping metals

There are several other ways of shaping metals. Wires are made from a rod of metal which is heated then cooled to soften the metal. Then the rod is cleaned, and a point is made at one end. The point is put into the hole in a die. The diameter of the hole in the die is slightly smaller than the diameter of the rod. The rod is gripped then pulled through the hole, so a narrower rod is formed (Figure 4.3). This will be repeated 8 or 9 times: each time the hole in the die will be smaller, until the desired thickness of wire has been made. Tubes can be made in two ways, as shown in Figures 4.4 and 4.5.

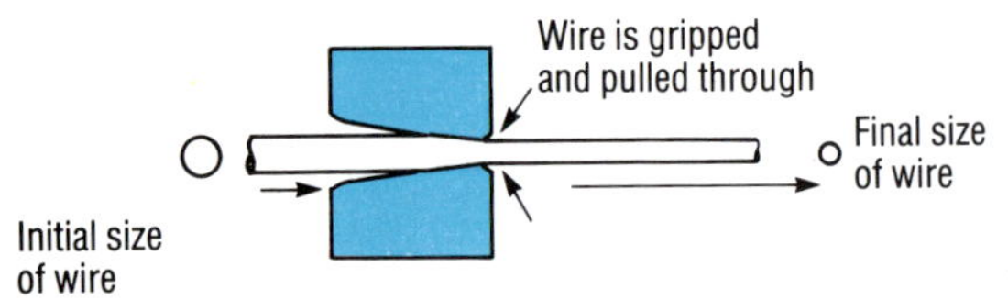

Figure 4.3 Wire drawing

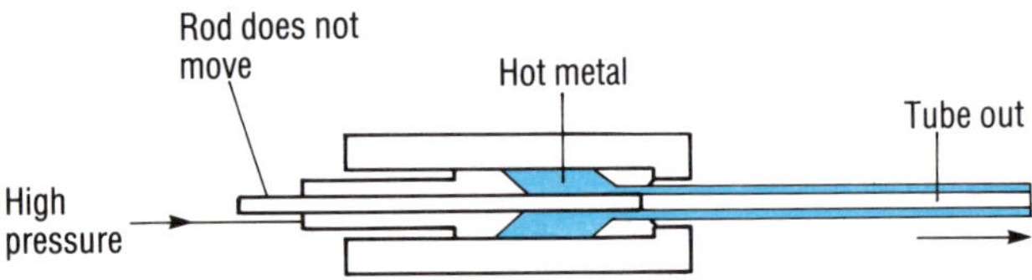

Figure 4.4 Shaping metal into a tube

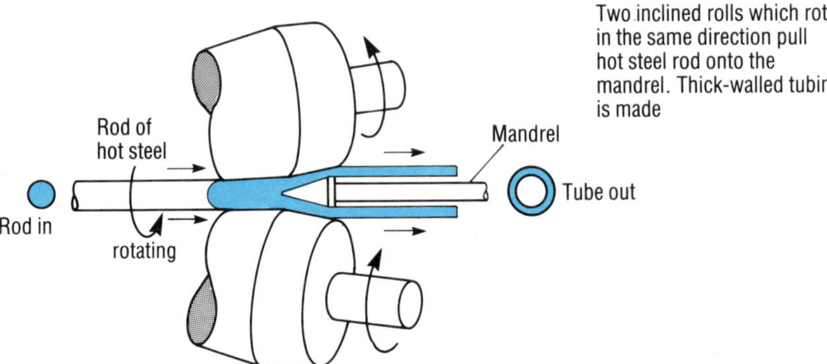

Figure 4.5 An alternative way of shaping metal tubes

Larger tubes can be made from a sheet of metal which is rolled up. The edges are brought together and then welded or soldered to seal the join.

Sheets of metal are made by repeatedly rolling metal strips between a series of rollers (Figure 4.6). The rollers apply a great pressure on the strip of metal. The strip of metal goes backwards and forwards through the rollers under tension. **Hot rolling** is used for a quick reduction in the thickness of a strip of metal. **Cold rolling** is usually used to make smooth and accurate thin sheets. A cold-rolling bay for aluminium strip is shown in Figure 4.7.

Lathes are also often used to work metal.

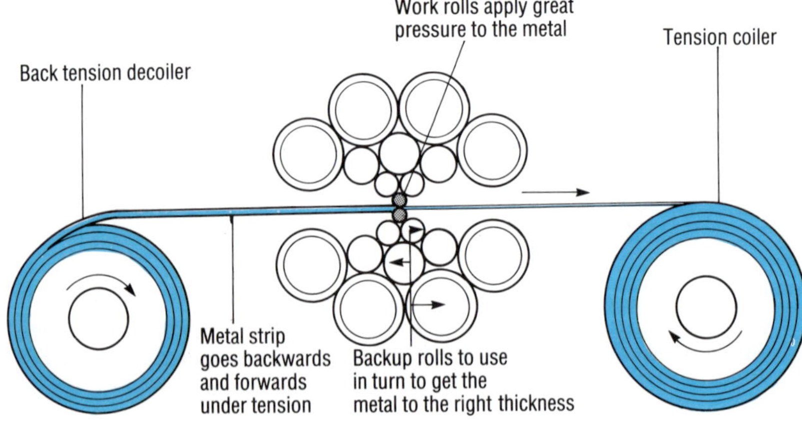

Figure 4.6 Rolling metal to make sheet

Figure 4.7 Cold-rolling bay for aluminium strip

Heat treatment of metals

In Unit 3, we saw that alloying can make a big difference to the way a metal behaves. **Heat treatment** is another important way of altering the properties of a metal so that it becomes more suitable for a particular job. Heating and then cooling the metal can be done in different ways (eg rapid or slow cooling).

In Activity 4.1 different forms of heat treatment cause the steel to behave in different ways. The heat treatment causes changes in the crystal structure of the iron in the steel, and it is these changes which alter the properties of the steel.

The hot Bunsen flame heats the steel to about 900° C. At this temperature, the internal crystal structure of the iron changes. If the steel is allowed to cool slowly, the structure simply changes back to its original form. If the steel is cooled suddenly, for example in cold water, the crystal structure is 'frozen' in the high-temperature form. This is called **quenching**. The steel is now hard, but brittle. If the steel is then re-heated to a lower temperature, and allowed to cool slowly, the steel remains hard, but loses its brittleness. This is called **tempering**.

Figure 4.8 Red-hot steel coming from the billet caster will then be quenched in cold water to give it a greater hardness

Activity 4.1

The heat treatment of steel

1 Collect four steel needles (ordinary darning needles).
2 Using pliers, try bending one of the needles. Which of these words would you use to describe the needle? You might find more than one appropriate. Is it hard, bendable, springy, tough, or brittle?
3 Heat one of the needles in a hot Bunsen flame until it is red hot, and leave it to cool slowly on a heat-proof mat.
4 Heat the remaining two needles until they are red hot, and drop them in a beaker of cold water.
5 Take one of the needles from the beaker, and heat it until it is blue (a much lower temperature). Then allow it to cool slowly on a heat-proof mat.
6 Using pliers, try bending the three heat-treated needles. What effect has heat treatment had on the steel?

Why food cans?

In 1795 Napoleon was worried about how his large armies could be fed. So he offered a reward for the invention of a practical method to preserve food. Nicholas Appert, a French chef, put food into stoppered bottles, which he heated in boiling water. He won the reward of 12 000 francs for his method, which involved similar stages of preserving food to those used in the canning process today.

How food cans are made

Most food cans are made by the three-piece method. The most recent ones are welded rather than soldered together because the lead in the solder was thought to be a health hazard.

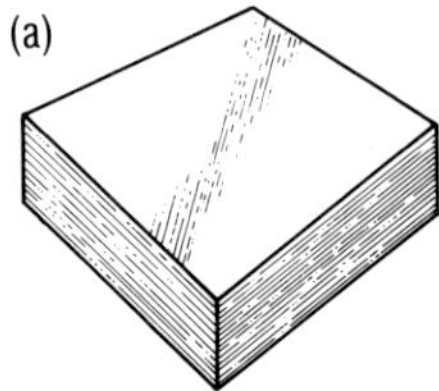

1 At the start of the processing line sheets of tin-plated steel are stacked. Some of the sheets may be coated with a barrier coating/lacquer. The lacquer is used to prevent direct contact between the food and the tinplate.

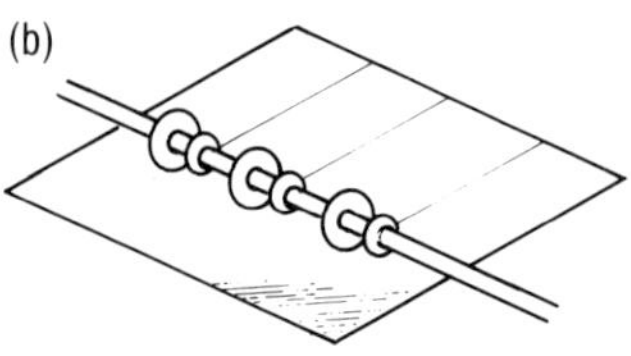

2 The sheets are cut into long strips, then into the widths required for specific can sizes.

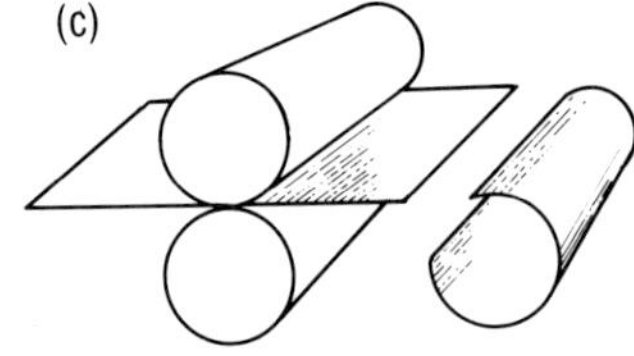

3 The individual strips of tinplate are rolled into cylinders.

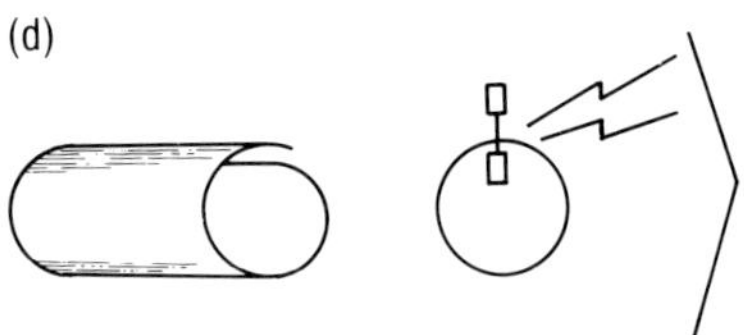

4 The edges of the cylinder are drawn together so they overlap. They are then welded electrically.

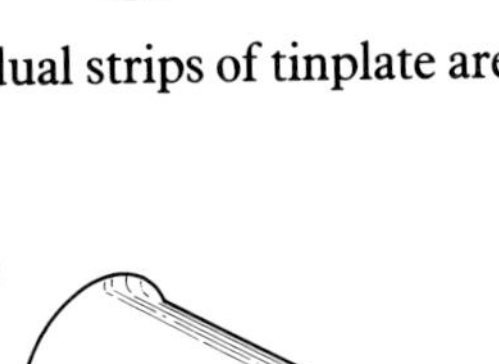

5 A lip is formed at each end of the cylinder.

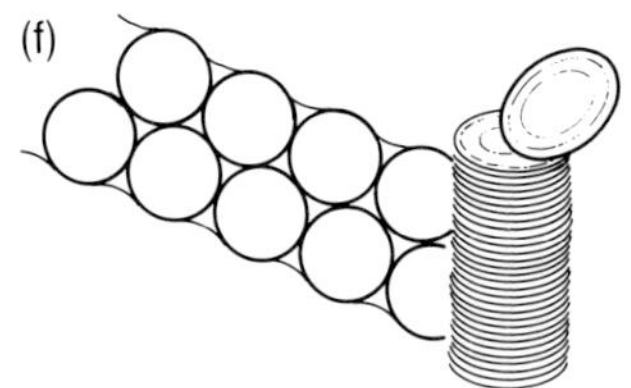

6 The ends for the cylinders are made in another part of the factory.

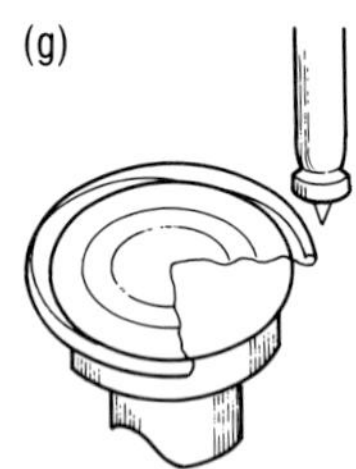

7 The rims of the ends are curled over.

8 The base is joined onto the cylinder and sealed with a sealing compound to form an airtight seal. The cans are ready to be sent, with the lids, to the food manufacturer to be filled and then sealed.

The canning process

Food is added to the open can. A liquid such as common salt solution (brine) for vegetables or syrup for fruit, is added. Then the lid is put on the can and sealed, so it is airtight. Then the cans are put into a large pressure cooker and heated. The food is cooked and sterilized by this heating. Finally the cans are cooled slowly.

Summary

Many metal objects are **cast** in **sand moulds**.
Metal dies are also used for casting objects, the advantage is that the same die can be used repeatedly.
Metal wire is made by pulling a softened metal rod through a hole in a die.
Metal tubes can be made by pushing a metal rod through a die or rotating a hot metal rod through towards a pointed mandrel.
Sheet metal is made by repeatedly rolling metal strip between high-pressure rollers.
Heat treatment changes the structure of metals and thus alters the properties. **Quenching** is heating a metal then cooling it suddenly; it makes metals harder. **Tempering** is heating a metal and allowing it to cool slowly; it makes metal less brittle.
Canning food allows a much longer shelf-life for the food. Food and liquid are sealed in the can to keep the air out. The food is cooked and sterilised.

Questions

1 What are the disadvantages of sand casting when compared with die casting?
2 Outline three different ways to make a metal tube.
3 Find out some other examples of uses of heat-treated steels. For each one state what heat treatment was used, what property the heat treatment gave the steel, and why that property is needed in that particular application.
4 Heat treatment is also used on other metals. Try to find out about some examples.
5 In the experiment on the heat treatment of steels (Activity 4.1), we could draw a sketch graph of temperature changes against time for each heat treatment. The graph for heating to 900° C, and then allowing to cool slowly, would look like this:

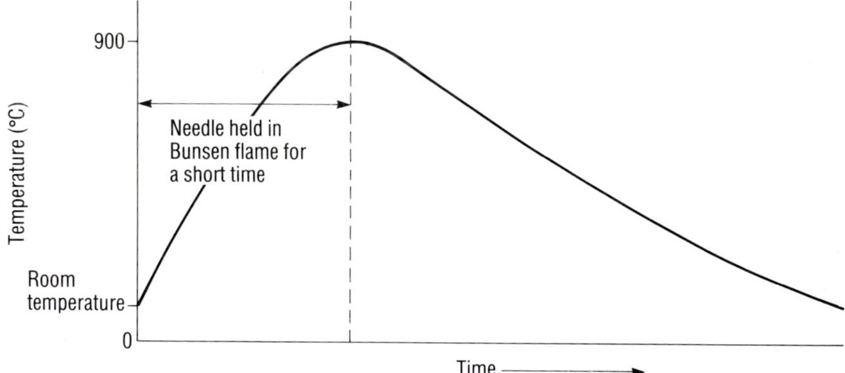

Draw sketch graphs for the other two kinds of heat treatment given to the steel in the experiment.

Activity 4.2

1 Look at tinned food cans at home. Try to see whether they are made by the three-piece method or the two-piece method, where all but the top is made from a single piece of metal.
2 Sketch the different shapes of tins of food. Suggest reasons why the food industry uses differently shaped cans.
3 Why should tinned foods be (a) airtight (b) heated in the canning process?
4 Why should you not eat food from a dented or bulging can?
5 The manufacturers recommend that food should not be stored in an opened can but should be stored in another container in the fridge. Why do they suggest this?
6 Look at the labels on some canned foods. Make a note of any artificial preservatives used. Compare your findings with those for non-tinned foods, like those in jars or packets. What conclusions can you draw?
7 List all the advantages of canned foods. Are there any disadvantages?
8 Nicholas Appert stored foodstuffs in glass containers. Why are cans preferred to glass today?

Unit 5
Why are metals so different from other materials?

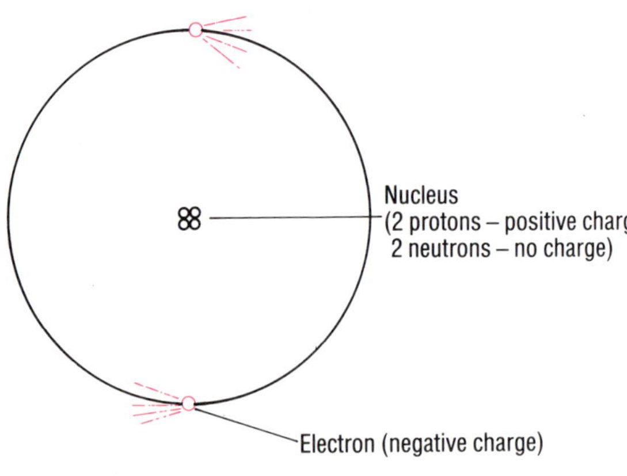

Atomic structure

The work of famous scientists such as Rutherford, Bohr, Chadwick and Thomson in the the nineteenth and early twentieth centuries suggested that all matter is composed of **atoms**.

They worked out that an atom consists of three small particles:

- protons
- neutrons
- electrons

Protons and **neutrons** have about the same mass. They are locked together in the middle of the atom, forming the **nucleus**. Protons have a positive electrical charge. Neutrons have no electrical charge.

Electrons are about 2000 times lighter than protons and neutrons. They occupy the outer part of the atom, moving round the nucleus at very high speed. Electrons have a negative electrical charge. As 'opposite charges attract' the negative electrons are held in orbit by the attraction of the positive protons.

The relative sizes and charges of the particles are summarised in the table below.

Particle	Relative mass	Charge
Proton	1	+1
Neutron	1	0
Electron	1/1837	−1

An atom is electrically neutral. This is because the number of electrons is always the same as the number of protons, so that the charges balance each other.

Elements

An **element** is a pure substance which cannot be split up by chemical reactions. There are 92 elements occurring naturally on Earth, and every substance that exists is made from these elements, put together in millions of different ways. A further 13 elements have been made artificially in laboratories. Most of the elements are metals – 79 of them. The remaining 26 are called non-metals.

An element consists of atoms of one type, all the same as each other. The atoms of one element are different from the atoms of every other element. For example, a piece of copper consists entirely of copper atoms. A piece of aluminium consists entirely of aluminium atoms. Atoms of copper are different from atoms of aluminium.

What makes an atom of one element different from an atom of another element is the number of protons in the nucleus (which is the same as the number of electrons in a neutral atom). For any particular element, the number of protons is always the same – it cannot vary.

For example, the number of protons in an atom of copper is always 29 (Figure 5.2(a)). The number of protons in an atom of aluminium is always 13 (Figure 5.2(b)). The number of protons in an atom of an element is called the atomic number of the element. So the atomic number of copper is 29, and the atomic number of aluminium is 13.

Where are the metals in the Periodic Table?

The **Periodic Table** shows all known elements in increasing order of atomic number (number of protons in one atom of the element). Elements in a vertical column of the table are very similar and are called a chemical **group**. The Roman numbers along the top of the table are group numbers. Only the elements in the top right-hand corner, and hydrogen, are non-metals.

Look at the Periodic Table (Figure 5.3) and answer the following questions:

1. Whereabouts in the table are
   (a) the most reactive metals?
   (b) the less reactive metals?
2. Give the letters (symbols) of the most reactive metal.
3. Give the general name for the biggest section of metals in the Periodic Table.
4. Give the group number for a chemical group of elements containing an equal number of metals and non-metals.
5. Give the group numbers for two groups containing six metals and only metals.
6. Which group or area of the table contains (a) Copper (Cu), (b) Zinc (Zn), (c) Aluminium (Al), (d) Lead (Pb), (e) Silver (Ag)?

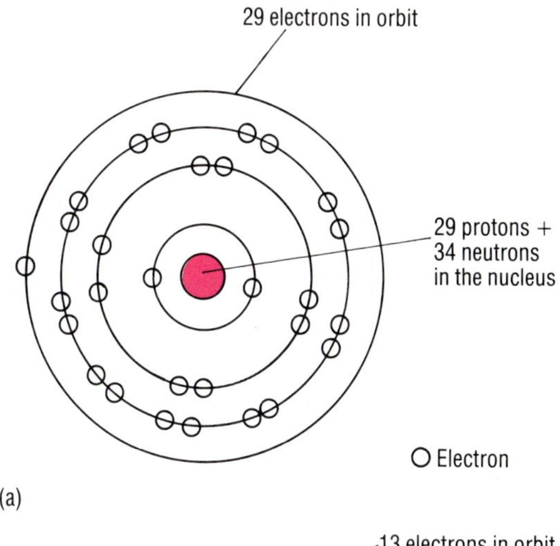

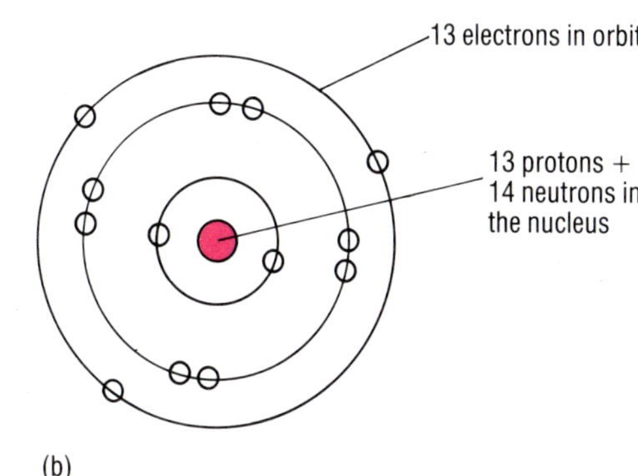

Figure 5.2 (a) An atom of copper; (b) An atom of aluminium

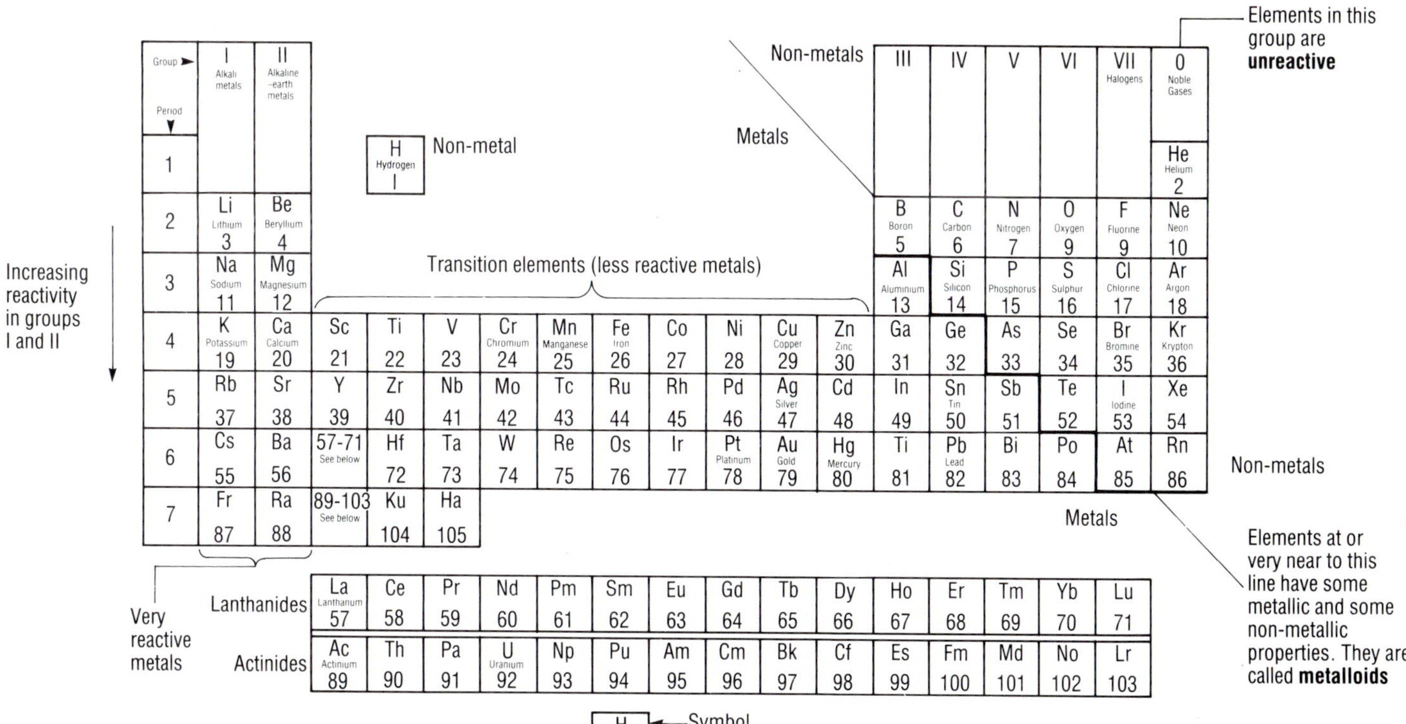

Figure 5.3 The pattern of chemistry – the Periodic Table

Isotopes

The number of neutrons in an atom can vary, but it still remains an atom of the same element. Atoms of an element with different numbers of neutrons are called isotopes. Most elements have isotopes. The mass number of an isotope is the total number of protons and neutrons in the nucleus. The mass number is used to identify a particular isotope. For example, uranium-235 is the isotope of uranium with mass number 235. Uranium has an atomic number 92. Therefore, uranium-235 has 143 neutrons in the nucleus (235 − 92).

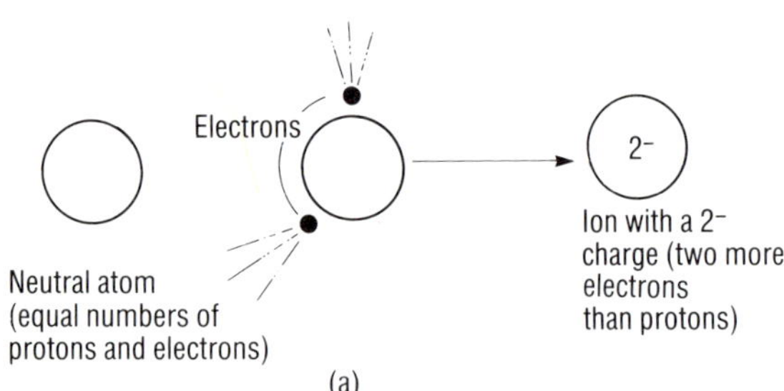

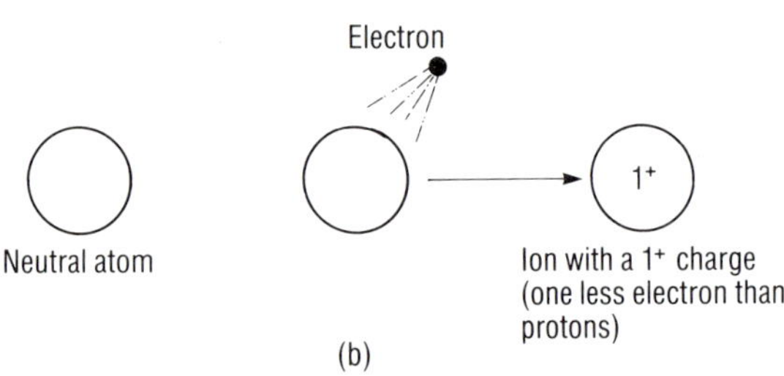

Figure 5.4 (a) A neutral atom gains two electrons to become a negative ion; (b) A neutral atom loses an electron to become a positive ion

What are ions?

As we have already seen atoms are electrically neutral because they contain equal numbers of protons and electrons, which balance each other. Since molecules are made of atoms, it follows that molecules are also electrically neutral.

Atoms and groups of atoms can lose or gain electrons, which means they are no longer neutral. The maximum number that can be lost or gained is *usually* three.

If a neutral atom gains electrons, it will have more negative charges. So it will become a negative ion (Figure 5.4(a)).

If a neutral atom loses electrons, it will have fewer negative charges than it had when it was a neutral atom. So it will become a positive ion (Figure 5.4(b)).

Metals always form positively-charged ions, if they form ions at all. This is because metal atoms have a larger radius than non-metal atoms. The outer electrons of a metal atom are thus further away from the positively-charged nucleus and are held less strongly. The outer electrons can be **lost** more easily, with the formation of a positive ion. Less energy is involved in losing a small number of electrons (typical of a metal) than gaining 5, 6 or 7 electrons to obtain a noble gas arrangement.

Metals react by losing their outer electrons (Figure 5.5). Also metals are different from non-metals as they have only one, two or three electrons in their outer orbit.

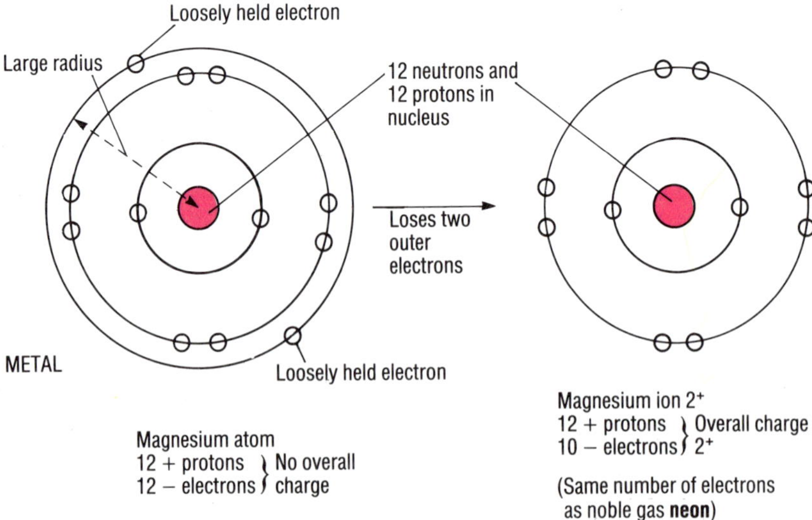

Figure 5.5 Magnesium atom loses electrons to become a positive magnesium ion

By contrast non-metal atoms have a smaller radius and do not lose electrons in chemical reactions. The nucleus has such a strong pull on the outer electrons in a non-metal that the atoms **gain** electrons to become ions (Figure 5.6).

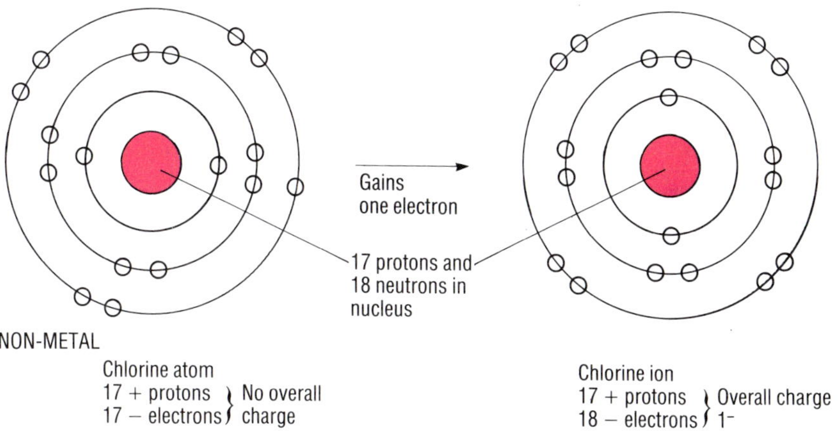

Figure 5.6 Chlorine atom gains an electron to become a negative chlorine ion

In Activity 5.1, you probably found that the purple colour from the potassium manganate (VII) moved towards the crocodile clip connected to the positive terminal of the lab pack. The purple colour is due to the manganate (VII) part of the potassium manganate (VII). The potassium part is colourless, and so you could not see it. If you could, you would have found it moving towards the negative terminal. Since positive and negative charges attract each other, this means that the potassium must be positively charged, and the manganate (VII) negatively charged. So the potassium and the manganate (VII) are not atoms or molecules, which are always electrically neutral. They are **ions**.

What holds a lump of metal together?

Solid metals contain positive ions of the metal in a regular arrangement in three dimensions (a lattice). In this solid lattice the outer electrons move about freely; they are not attached to a particular ion (Figure 5.7). This 'sea' of moving outer electrons is like a negatively-charged net holding the positive ions in place; opposite charges attract. The high melting point of metals shows that the attractive forces in a metal lattice are large.

The outer electrons move about in all directions, holding the positive ions in place. In the metal lattice the positive ions actually push against each other. The positive metal ions only vibrate about a fixed position in the lattice.

The properties of metals

The theory of how metals are bonded can explain the special properties of metals. They are strong, malleable and ductile because if a stress is applied to a metal the sea of electrons is still there to hold the structure together and pull it back into shape (Figure 5.8).

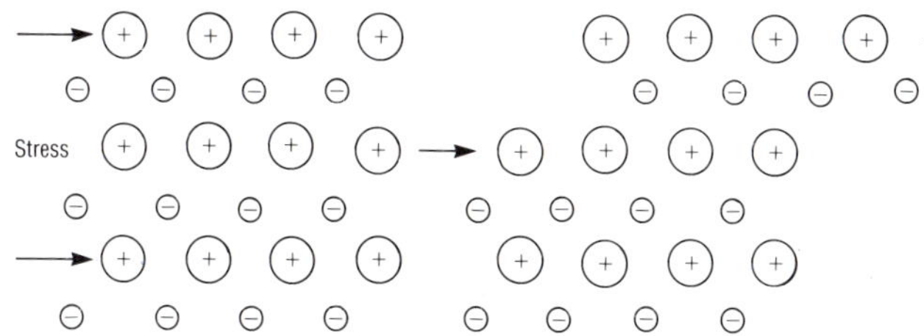

Figure 5.8 If a stress is applied to a metal (eg the metal is stretched), the sea of electrons holds the structure together and the forces of attraction pull the metal back into shape

Activity 5.1

What happens to potassium manganate (VII) in an electric field?

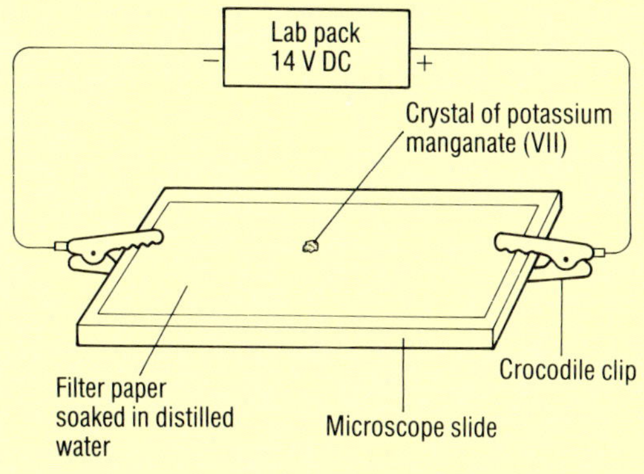

1. Cut a piece of filter paper so that it just fits on a microscope slide.
2. Soak the filter paper in distilled water, and place it on the slide.
3. Place a crystal of potassium manganate (VII) in the centre of the filter paper.
4. Connect the ends of the filter paper to a lab pack using crocodile clips.
5. Switch on.
6. Observe the purple colour from the crystal which soaks into the filter paper. Which way does it move, towards the positive or towards the negative?

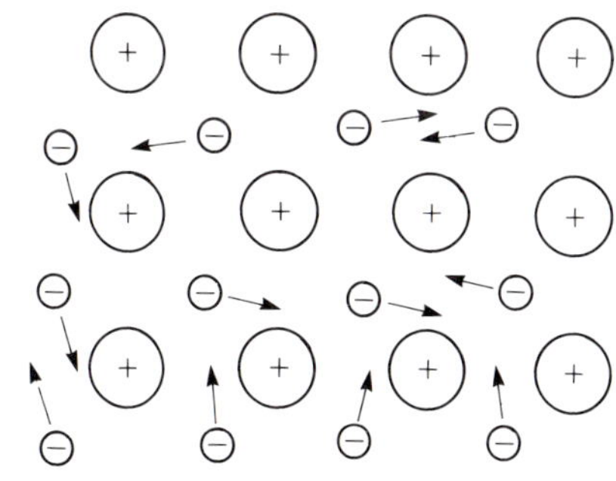

Figure 5.7 A metal lattice

Metals are good conductors of electricity because as soon as a voltage is placed across the metal the sea of electrons moves towards the positive terminal (opposite charges attract). The positive ions cannot move in the solid.

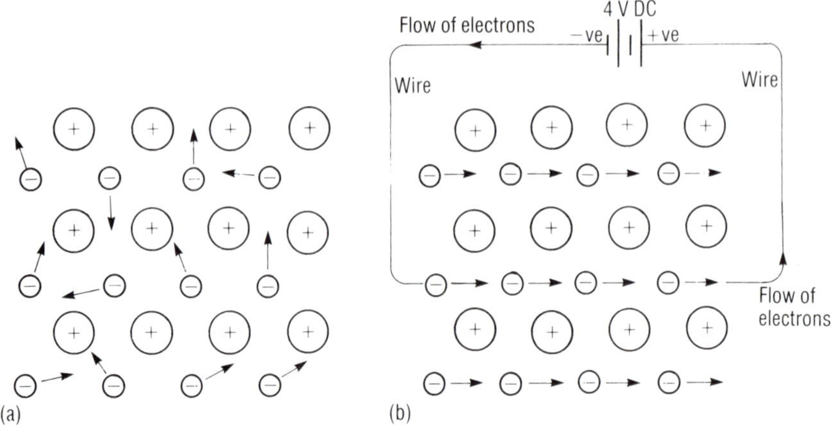

Figure 5.9 (a) Metal lattice before a voltage is applied; (b) After a voltage is placed across the metal – the electrons flow in one direction only and an electric current flows

Metals are good conductors of heat because the sea of electrons moves about (flows) and transmits heat. Conduction of heat is due to the fast movement of particles in a substance from one end to the other. Metals are solids with high melting points because the sea of electrons holds the lattice together so strongly; the metallic bond is very strong.

Non-metallic elements and compounds do not behave like metals because they do not have the sea of free electrons. Graphite is the only exception, it is a good conductor of heat.

Outward signs of the arrangement in metals

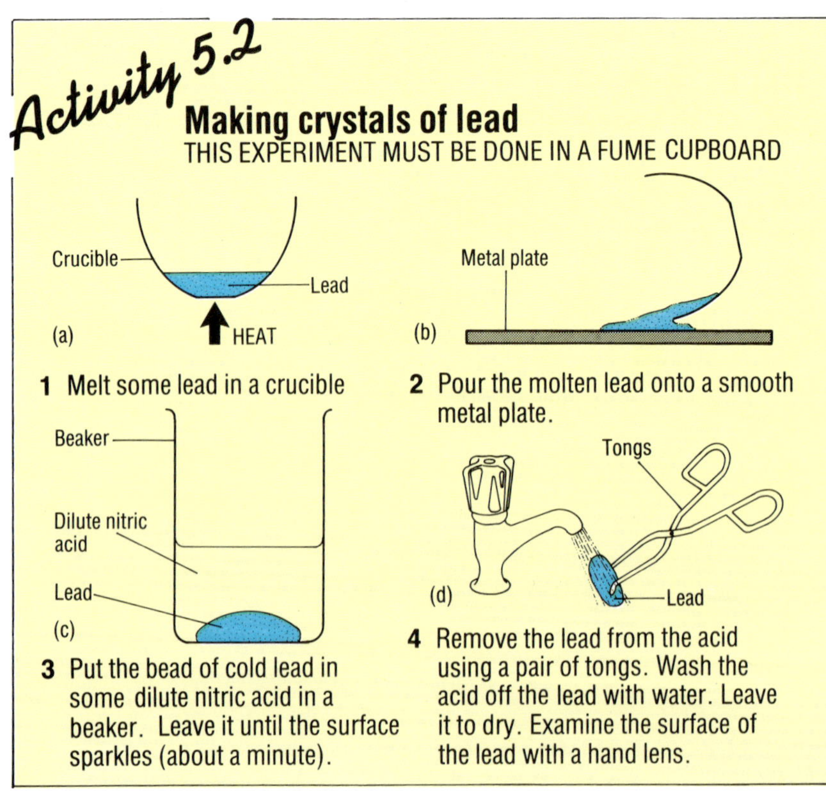

In Activity 5.2, three things have happened: the metal melted when heated, then it solidified when cooled. Crystalline areas, called grains, formed on its surface. They are of different shapes and sizes.

Starting from these pieces of evidence, we can produce a model (Figure 5.10) of what the metal might be like inside, which will explain this behaviour.

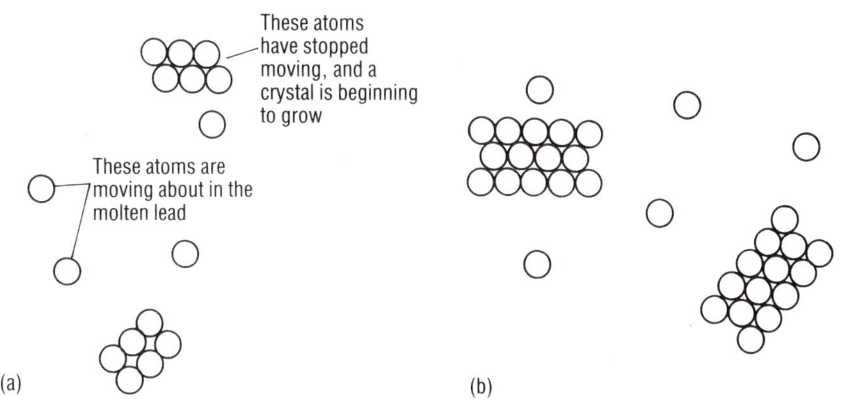

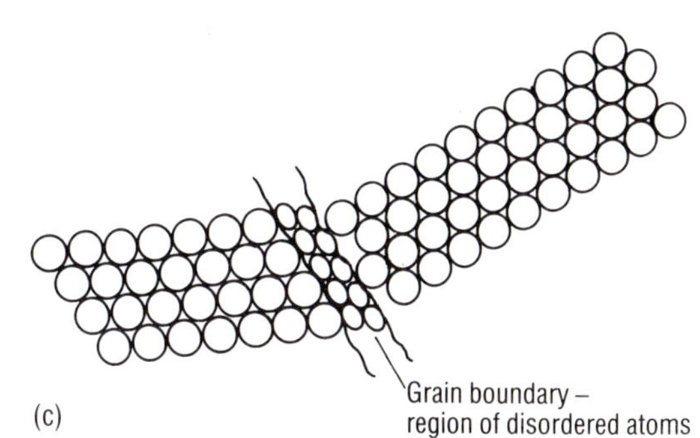

Figure 5.10 (a) The metal starts to cool; (b) Cooling continues – two obvious crystals are growing; (c) Crystals of different shapes are formed – the grain boundary is where the crystals meet

Activity 5.3

A bubble raft model of a metal structure

This experiment produces a two-dimensional model of the internal structure of a metal.

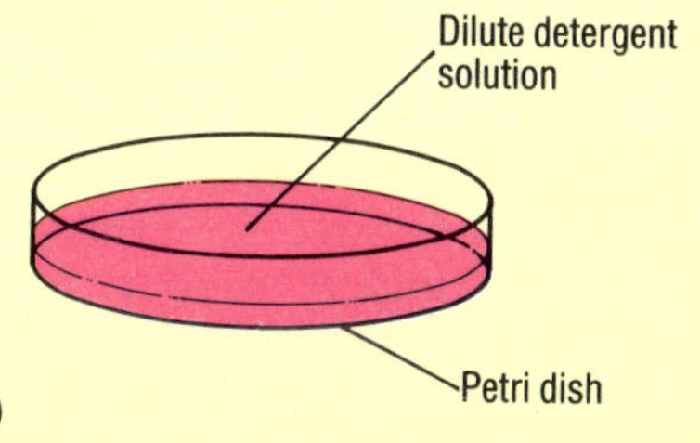

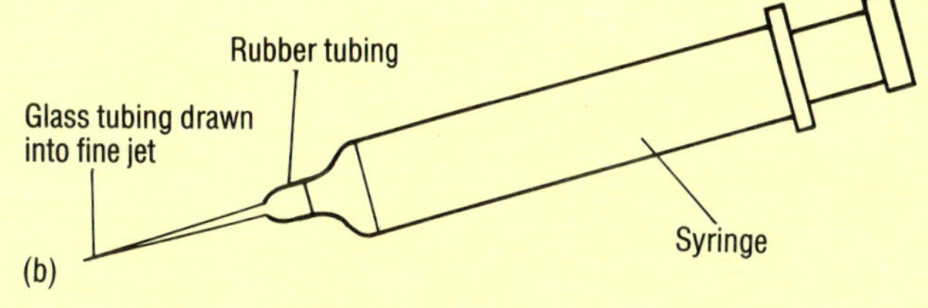

1 Put some dilute detergent solution in a petri dish: 10 cm^3 of teepol in 1 litre of distilled water is suitable.

2 Attach a piece of glass tubing, drawn into a fine jet, to the end of a syringe using rubber tubing.

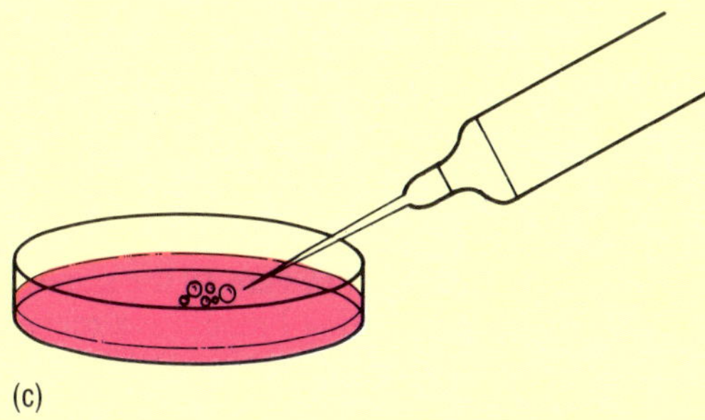

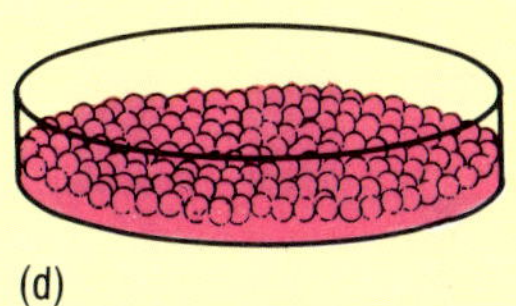

3 Place the end of the glass tubing under the surface of the detergent solution. Squeeze the syringe gently to produce bubbles of equal sizes (about 2 or 3 mm across).

4 Move the syringe about slowly, until the whole surface of the detergent solution is covered in bubbles. Examine the bubble pattern produced.

Notes on the experiment: In your bubble raft, you should be able to find areas where the bubbles are arranged regularly. These areas represent **grains** in a real metal. Can you find any grain boundaries?

Look for places where a row of bubbles is displaced, or comes to an end. This is called a **dislocation**. It may be caused by a large bubble, which represents an atom of impurity in a real metal. A dislocation makes the metal much weaker. Depending on what has happened to the metal previously, there could be millions of dislocations in a fairly small piece of metal.

Summary

Metals, in common with all substances, contain **atoms**.
Atoms contain **protons, electrons** and **neutrons**.
The number of electrons in a neutral atom always equals the number of protons.
Metals make up the great majority of known **elements**.
Metals usually **lose** outer electrons to make **positive ions**.
Non-metals usually **gain** outer electrons to make **negative ions**.
A lump of metal consists of positive ions held together by a 'sea' of electrons. The 'sea' of electrons is the key to the behaviour of metals.
A bubble raft shows the arrangement of **grains** and **dislocations** found in solid metals.

Questions

1 Explain what is meant by an ion.
A magnesium atom becomes an ion by losing two electrons. What charge will it have? How does an aluminium atom become an ion with a charge of $3+$?

2 Explain how a piece of aluminium conducts electricity.

3 What is meant by the atomic number of an element? Can two elements have the same atomic number? Explain your answer.

4 Explain what is meant by the word isotope. Copper has two isotopes, copper-63 and copper-65. The atomic number of copper is 29.
For each isotope, find (i) the number of protons in one atom, (ii) the mass number, (iii) the number of neutrons in one atom.

5 Aluminium has only one isotope. Its atomic number is 13 and its mass number is 27. Find (i) the number of protons in one atom, (ii) the number of neutrons in one atom, (iii) the number of electrons in one atom.

6 Look around at school, at home, or on the street for examples of grains in metal surfaces. Make a note of any you find. If you have a camera, you could take photographs of the examples you find, and put them in your notes. Try to find out what the metal is in each case.

7 Explain the arrangement between positive ions in a metal lattice and the 'sea' of electrons.
How does this arrangement explain the main properties of metals?

8 Why don't all other substances/elements have metallic properties?

Crossword on Unit 5

Clues across

1 Smallest part of atom
4 ...-made fibres can become electrically charged
6 Positive particle
7 It is made of a lattice of atoms
8 ...... Rutherford, scientist
10 Sea-creature that can be electric
12 1 down is a ...... substance
14 You might cook in it using 2 down
15 Charge on 1 across

The solution is on page 62

Clues down

1 Its atoms are all the same
2 Moving 1 across a in metal produces this
3 Not 7 across!
4 The stuff of science
5 Centre of atom
9 The size of the positive charge on an aluminium ion
11 You might cook it in 14 across
13 Formed when an atom gains or loses electrons
14 Number of isotopes of aluminium

Unit 6
Metals reacting with oxygen

Activity 6.1

Reacting metals with oxygen

Wear safety goggles throughout this experiment.
Your teacher will provide you with several boiling tubes full of oxygen.

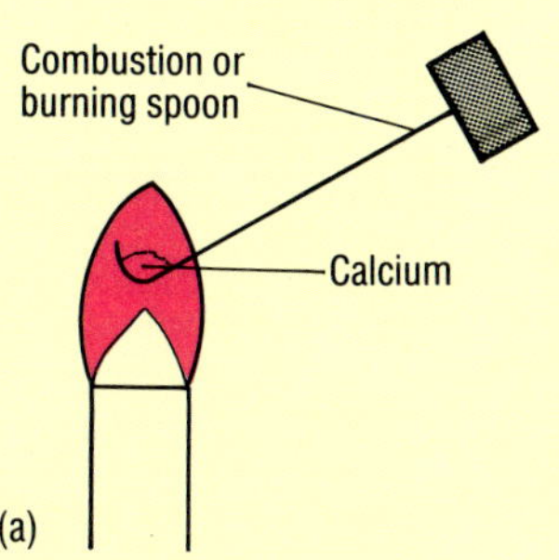

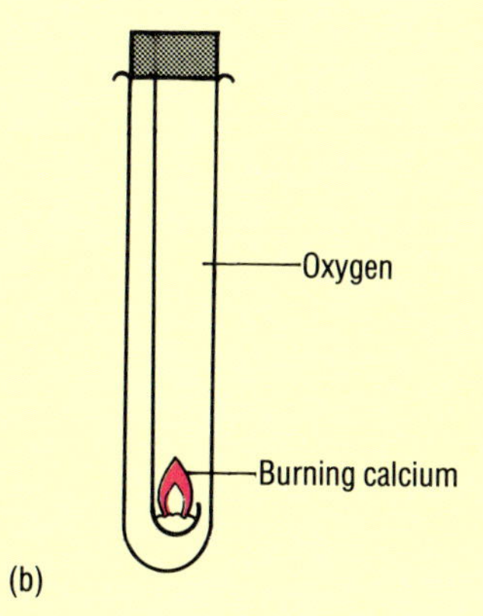

1 Place a piece of calcium on a burning spoon. Heat it strongly in a roaring blue Bunsen flame until it starts to burn.

2 When the calcium starts to burn, put the spoon into a boiling tube of oxygen.

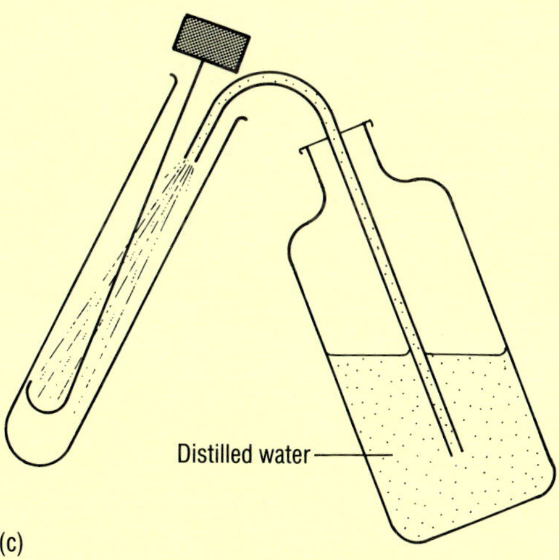

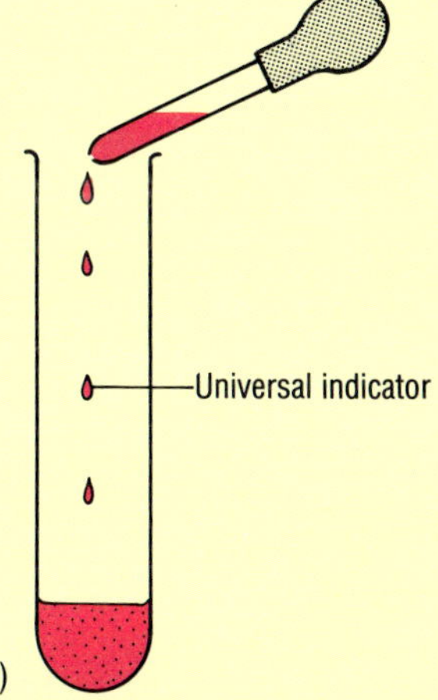

3 When the calcium stops burning, add distilled water to the ash to try to dissolve it.

4 Test any solution formed with universal indicator.

5 Repeat using iron filings, copper powder and zinc powder. Your teacher may supply magnesium ribbon which should be held in tongs (**DO NOT LOOK AT THE BURNING MAGNESIUM**). You may also be shown sodium burning in oxygen.

6 Make a list of the metals in order of how well they react with oxygen, the most reactive first.

Burning

When metals burn in oxygen (including oxygen in the air) they combine with oxygen to form oxides.

For example

$$\text{magnesium} + \text{oxygen} \rightarrow \text{magnesium oxide}$$
$$2\,Mg + O_2 \rightarrow 2MgO$$

(The oxide formed when sodium burns is a little unusual. It is called sodium peroxide.) See Activity 6.1 for some other examples.

If the oxide of a metal dissolves in water, it forms an alkaline solution.

Reactivity of metals with oxygen

From the results obtained in Activity 6.1 we can see that some metals are more reactive with oxygen than others. It is not possible to put the metals in a definite order, from these results alone, but we can group them as follows.

- metals that burn: magnesium, sodium, calcium
- metals that glow: zinc, iron
- metals that hardly react: copper

This order of reactivity is a measure of how strongly the metals want to combine with oxygen.

Competing for oxygen

If some metals combine with oxygen more strongly than others, what happens when there is competition for oxygen? (Activity 6.2 should help to answer this question.)

Activity 6.2

Metals competing for oxygen

In this experiment iron and copper compete for oxygen. From what you learned in Activity 6.1, which metal would you expect to win the competition?

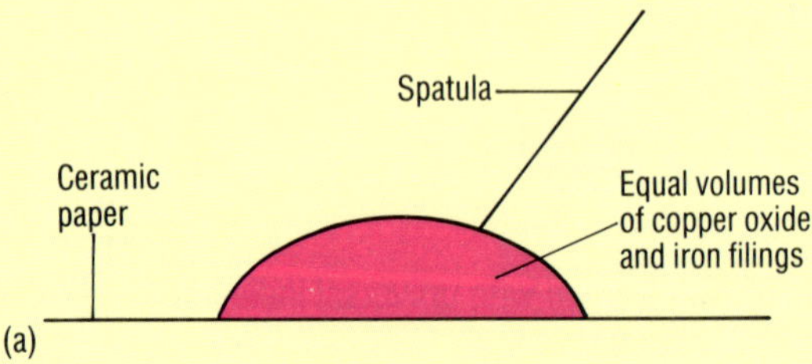

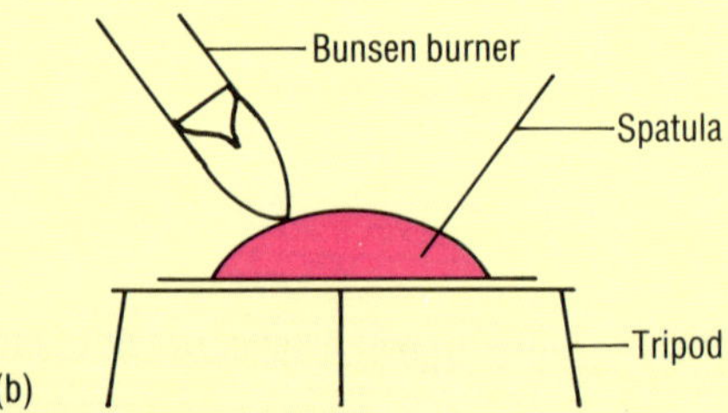

1. Mix together a little copper oxide and an equal volume of iron filings on ceramic paper. Note the colour of the mixture.
2. Place the paper on a tripod. Heat the mixture strongly from above with a Bunsen burner, while stirring with a spatula.
3. If you see a glow spread through the mixture, it means a reaction has occurred. Stop heating, and examine the mixture.
4. If you do not see a glow spread through the mixture, heat for 5 minutes, and then examine the mixture.
5. What colour is the mixture? What has been formed? Does this support your prediction?
6. Repeat the experiment using iron filings and zinc oxide, and then iron filings and lead oxide.

In a competition reaction, a more reactive metal takes oxygen away from a less reactive one. For example

$$\text{aluminium} + \text{iron(III) oxide} \rightarrow \text{iron} + \text{aluminium oxide}$$
$$2Al + Fe_2O_3 \rightarrow 2Fe + Al_2O_3$$

Aluminium takes oxygen away from iron, because aluminium is more reactive than iron. Aluminium wins the competition for oxygen.

The Thermit reaction

When one substance takes oxygen away from another the process is called reduction. When metals such as iron were first obtained from their ores, it was noted that the amount of iron produced was much less than the amount of iron ore from which it was obtained. In other words the iron ore had been reduced.

The Thermit reaction is an example of reduction. In this case aluminium reduces iron(III) oxide to iron. 'Thermit' is a historical term which we still use today.

The Thermit reaction is used to weld the ends of railway lines together (Figure 6.1). This gives passengers a smoother ride. The molten iron produced in the reaction is run onto the joint between the lines.

Reactivity with oxygen

This table compares the reactions of various metals with oxygen. It also describes the appearance of the oxides of these metals, the solubility of their oxides in water and the pH of the solutions formed when these oxides are dissolved in water.

Metal	Reaction with oxygen	Appearance of oxide	Solubility of oxide in water	pH of oxide solution
Calcium	Burns with red flame	White powder	Fairly soluble	10
Magnesium	Burns with bright white flame	White powder	Slightly soluble	9
Sodium	Burns with yellow flame	White solid	Soluble	12
Copper	Does not burn, turns black	Black solid	Insoluble	
Iron	Glows red and sparkles	Black solid	Insoluble	
Zinc	Does not burn, yellow coating formed	Yellow coating turns white when cool	Insoluble	

From the results of the experiments you should be able to write down an accurate order of reactivity of metals with oxygen. Remember that in a

Activity 6.3

A more vigorous competition reaction

Your teacher will demonstrate this reaction.

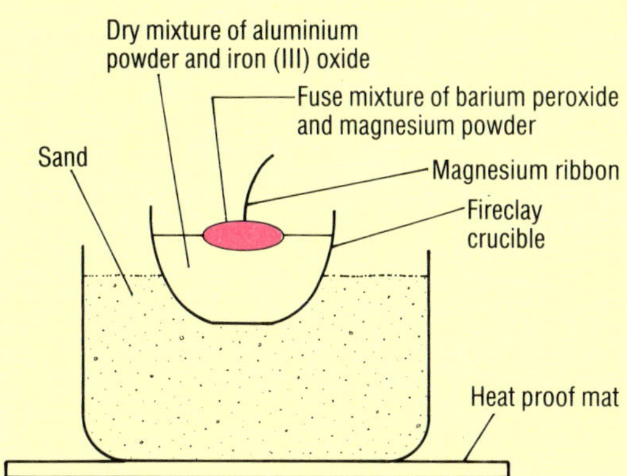

The magnesium ribbon is lit, and this lights the fuse mixture. This in turn provides the energy to start the reaction between the aluminium and iron(III) oxide. At the end of the experiment, iron is left. This reaction is known as the **Thermit reaction**.

Figure 6.1 The Thermit reaction is used to weld lengths of railway track into a smooth continuous length. This gives greater passenger comfort

competition for oxygen the metal higher in the list will win. A higher metal will take oxygen away from a metal lower in the reactivity order.

- Does your list agree with the one below?

Order of reactivity of metals with oxygen:

sodium }
calcium } difficult to decide the order
magnesium }
aluminium
zinc
iron
lead
copper

Summary

Metals combine with oxygen readily to make bases.
Some bases dissolve in water to give an alkaline pH.
More reactive metals will take oxygen from less reactive ones on heating. This can be used to decide a list of metals in order of reactivity; (more reactive) potassium, sodium, calcium, magnesium, aluminium, zinc, iron, lead, copper (less reactive).

Questions

1 Potassium is a metal which needs great care in handling.
In the reactivity list we have obtained so far, it would be placed at the top.
Predict:
(i) How it will react with oxygen.
(ii) The appearance of the oxide.
(iii) Whether the oxide will be soluble.
(iv) The pH of the oxide solution, if formed.

2 Look at the following pairs of substances, and then answer these questions
a lead oxide and aluminium
b magnesium oxide and zinc
c copper oxide and magnesium
d calcium oxide and copper
In which cases do you expect a reaction to take place when the mixture is heated?
For those in which you expect a reaction to take place:
(i) Say why you expect a reaction to take place.
(ii) State the names of the substances formed.
(iii) What sort of reaction has taken place?

3 Several metals burn brightly in air, and give a bright light that might be useful in a photographic flash bulb or flash cubes. Of these metals, only one is really suitable. Which one? Explain your answer.

Unit 7
Metals reacting in aqueous solutions

Activity 7.1

Reacting metals with cold water

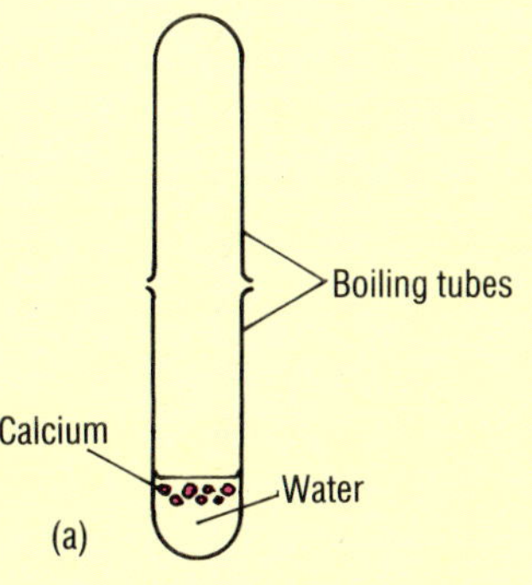

1 Put some distilled water in a boiling tube. Add a few granules of calcium, and place a second boiling tube over the first. Observe the reaction.

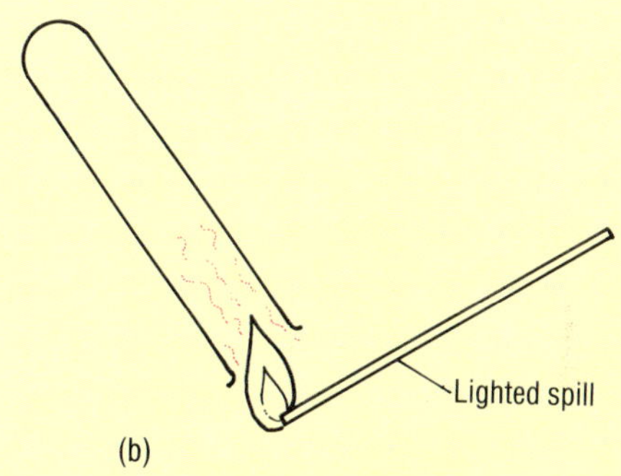

2 Remove the top boiling tube and test the gas with a lighted spill. What happens? What is the gas?

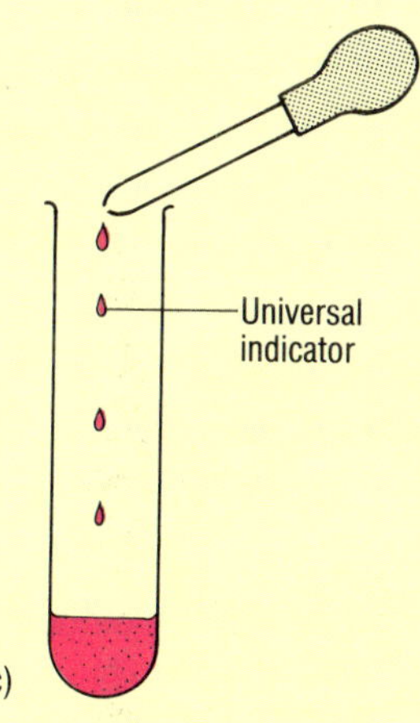

3 Test the solution with universal indicator. Is it acid, alkaline, or neutral?

Metal	Reaction
Calcium	
Copper	
Magnesium	
Iron	
Zinc	

4 Try reacting these metals with cold distilled water: copper, magnesium, iron, zinc. Look carefully for signs of reaction, especially bubbles of gas on the surface of the metal. Make a table like the one above.

Metals reacting in water

When calcium reacts with cold water, it does so steadily, giving off hydrogen gas. The hydrogen gas is displaced from (pushed out of) the water. A solution of calcium hydroxide is formed, which is an alkali.

Activity 7.2

Can some metals react with steam?

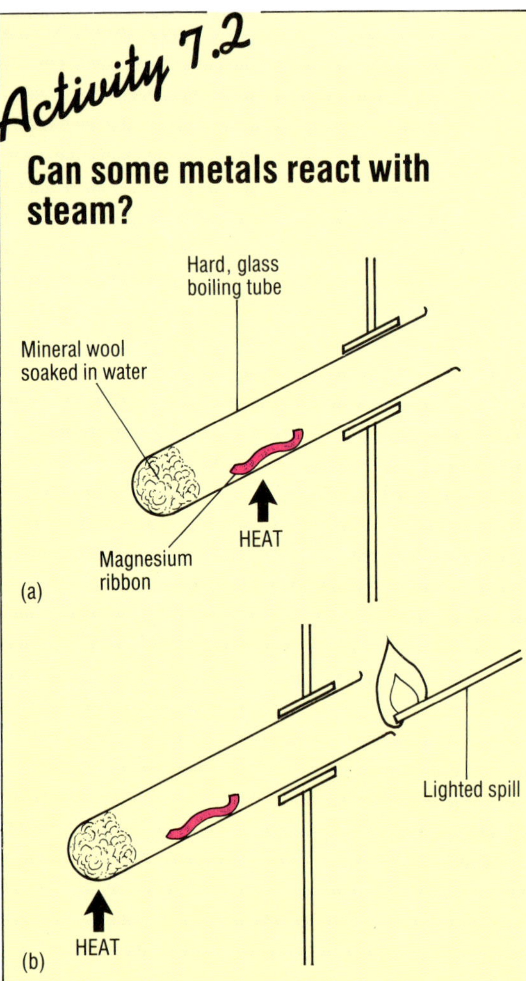

1. Heat a piece of magnesium strongly until it starts to burn (a).
2. Heat the mineral wool soaked in water to produce steam (b).
   Hold a lighted spill at the end of the glass tube. Is hydrogen given off?
3. Observe any changes in the metal. When the test tube is cool, carefully scrape out any ash. Try to dissolve it in water, and test any solution formed with universal indicator.
4. Repeat the experiment with zinc, iron and copper.

calcium + water → calcium hydroxide + hydrogen

$$Ca + 2H_2O \rightarrow Ca(OH)_2 + H_2$$

Your teacher will show you how two more metals, sodium and potassium, react with cold water. These metals are stored under oil. Why? Make notes as your teacher carries out the reactions.

Sodium reacts vigorously with the water. The heat energy produced by the reaction melts the sodium into a small ball. Another part of the energy produced by the reaction is used in making the sodium shoot around on the surface of the water. Hydrogen is given off, and sodium hydroxide, an alkali, is formed in the water.

sodium + water → sodium hydroxide + hydrogen

$$2Na + 2H_2O \rightarrow 2NaOH + H_2$$

The reaction with potassium is even more vigorous – it bursts into flames as it shoots around the surface. Again an alkali – potassium hydroxide – is formed in the water. Hydrogen is given off, but it is burned up by the flame.

potassium + water → potassium hydroxide + hydrogen

$$2K + 2H_2O \rightarrow 2KOH + H_2$$

In Activity 7.1, you probably found that the other metals you tried in your experiment did not seem to react with cold water. But you should have noticed that after a while, a few bubbles appeared on the surface of the magnesium. This suggests that a reaction may have been taking place very slowly, giving off a gas. Try to design an apparatus you could leave for several days which would allow you to collect and test the gas. Try the experiment. Write a word equation for any reaction that occurs.

Reactions with steam

Magnesium, which reacts very slowly with cold water, will react with steam if it is heated strongly. The magnesium burns, and reacts with the steam, displacing hydrogen from it. Magnesium oxide is left in the test tube, which will dissolve in water to form an alkali. What is the name of the alkali?

magnesium + steam → magnesium oxide + hydrogen

$$Mg + H_2O \rightarrow MgO + H_2$$

Displacement reactions

The reactions which occur in Activity 7.3 are **displacement** reactions. For example, you will find a reaction taking place between iron and copper(II) sulphate solution. The flaky pieces of metal you see in the solution are pieces of copper. The iron has displaced the copper from the copper(II) sulphate solution. The iron has gone into solution in its place.

iron + copper(II) sulphate → iron(II) sulphate + copper

$$Fe(s) + CuSO_4(aq) \rightarrow FeSO_4(aq) + Cu(s)$$

Activity 7.3

Displacement of metals

This experiment is a sort of competition among metals.

You will need clean pieces of iron, copper, tin, magnesium, lead and zinc. You will also need solutions of the nitrate or sulphates of these metals, and a solution of silver nitrate.

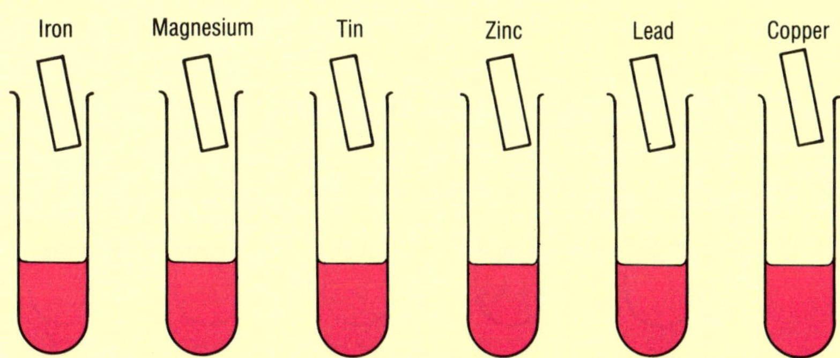

Copper (II) sulphate solution in each test tube

1. Put some copper(II) sulphate solution into six test tubes. Into the first put a piece of iron, into the second a piece of magnesium, tin in the third, zinc in the fourth, lead in the fifth and copper in the sixth.
2. Observe the test tubes. After two minutes, examine each test tube and each piece of metal for signs of a reaction. Look especially for pieces of flaky metal in the solution or on the pieces of metal. Record your observations.
3. Repeat the experiment using each of the other solutions in turn. Use fresh, clean pieces of metal each time. Do any reactions produce attractive metal crystals?
4. Record all your results in a table like the one below. Put a tick in the box if a reaction took place.

Metal	Magnesium solution	Zinc solution	Copper(II) solution	Tin(II) solution	Iron(II) solution	Lead(II) solution	Silver solution
Magnesium							
Zinc							
Copper							
Tin							
Iron							
Lead							

5. List the metals in order, with the one which did the most reactions first.

Reaction of metals with acids

When a metal reacts with an acid, hydrogen is displaced from the acid. (See Activity 7.4.) For example

$$\text{magnesium} + \text{sulphuric acid} \rightarrow \text{magnesium sulphate} + \text{hydrogen}$$
$$Mg + H_2SO_4 \rightarrow MgSO_4 + H_2$$

As well as displacing hydrogen, the metal forms a new compound with the acid, called a **salt**. When magnesium reacts with sulphuric acid, the salt formed is magnesium sulphate.

From the reactions observed in this unit, try to put the metals in order of reactivity. Compare it with the order found in Unit 6 (page 30).

Activity 7.4 Reaction of metals with dilute sulphuric acid

Wear safety goggles

Use magnesium, copper, zinc, lead, tin and iron in this experiment.

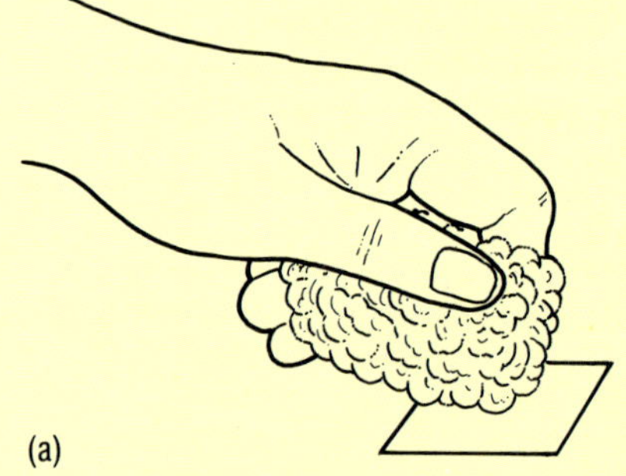
(a)

1 Clean each piece of metal with wire wool.

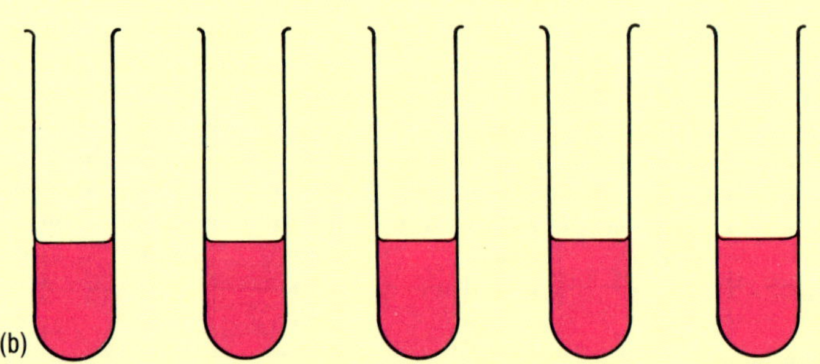
(b)

2 Put a few cm^3 of dilute sulphuric acid into each of the test tubes.

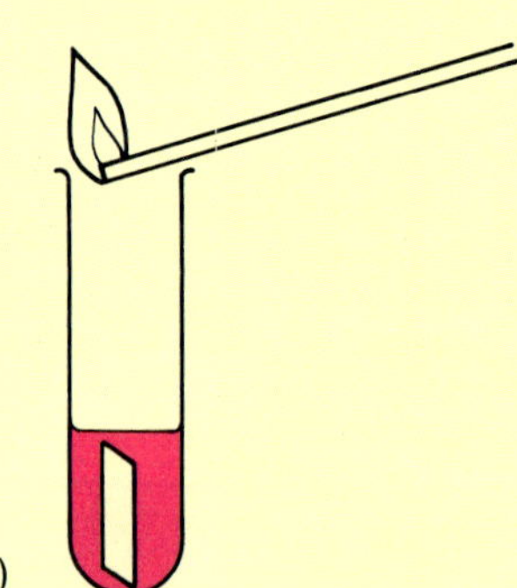
(c)

3 Put a piece of the first metal into the first test tube. If a gas is given off, test it with a lighted splint.

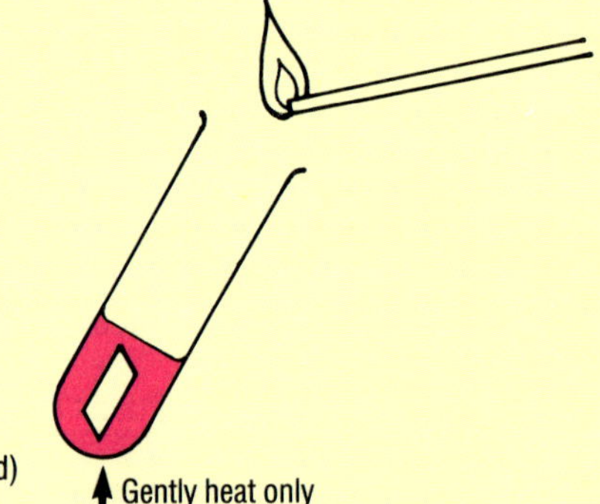

(d)

4 If there is no reaction, warm the test tube **gently** and test again for hydrogen.

5 Observe any other reactions.

6 Repeat with the other metals, using the remaining test tubes.

7 Record your observations in a table like the one below.

8 List the metals in order of reactivity with sulphuric acid.

Metal	Reaction with acid	Is hydrogen given off?

The special case of aluminium

Normally, aluminium is covered in a thin layer of its oxide. The oxide layer protects it, and prevents it from reacting. If this layer is broken, however, aluminium is a fairly reactive metal (see Activity 7.5).

Rubbing aluminium with mercury(II) chloride removes the oxide layer. Oxygen from the air reacts rapidly with the exposed aluminium, forming a whisker-like growth of aluminium oxide, and becoming hot. In sulphuric acid, aluminium which has been treated with mercury(II) chloride reacts strongly, giving off hydrogen. Where would you place it in the reactivity series?

Two metals not yet mentioned in this unit are silver and gold. One of the reasons they are made into fine jewellery is that they are very unreactive. Gold, especially, keeps its attractive shiny colour for centuries, and never tarnishes, as shown in Figure 7.1.

Platinum, another rare and valuable metal, is even more unreactive than gold or silver. These three metals are therefore at the bottom of the reactivity series.

As we have seen above, aluminium is a special case; it can be placed in the reactivity series between magnesium and zinc.

We can now produce a final version of the reactivity series.

(i) potassium
(ii) sodium
(iii) calcium
(iv) magnesium
(v) aluminium
(vi) zinc
(vii) iron
(viii) tin
(ix) lead
(x) copper
(xi) silver
(xii) gold
(xiii) platinum

One of the important tasks facing a scientist is to search for patterns and order in his/her study of science. The reactivity series provides a good example of how order is derived from a seemingly wide range of different reactions. In the following units, we will see how useful the reactivity series is in studying several other areas of metal behaviour.

Activity 7.5

How reactive is aluminium?

1. Try burning a piece of aluminium on a burning spoon.
2. Put a piece of aluminium in a test tube of sulphuric acid.
   - Does aluminium seem to be a reactive metal?
3. Immerse a strip of aluminium in mercury(II) chloride solution for about one minute. Remove the aluminium with tongs. Put it on a heat-proof mat, and observe it for a few minutes.
   (Warning: Do not touch the mercury(II) chloride; it is poisonous)
4. Repeat step **3**, then put the aluminium in a test tube of sulphuric acid. Test any gas given off with a lighted spill.
   - Have you changed your mind about the reactivity of aluminium?

Figure 7.1 This ancient gold bangle has kept its shiny colour for centuries

Reactivity of metals

Reactivity with water

Potassium: Catches fire in cold water. Forms potassium hydroxide, hydrogen given off (but it burns).
Sodium: Reacts vigorously with cold water. Forms sodium hydroxide, hydrogen given off.
Calcium: Reacts steadily with cold water. Calcium hydroxide formed, hydrogen given off.
Magnesium: Very slow reaction with cold water. Burns in steam to form magnesium oxide and hydrogen.
Zinc: Some reaction with steam, zinc oxide formed, hydrogen produced.
Iron: Little reaction with steam, iron oxide formed, little hydrogen given off.
Copper: No reaction with water or steam.
The order of reaction is the same as in the activity table of metals.

Displacement reactions

In Activity 7.3 metals were placed in solutions of metal salts. It was found that some metals displaced others from their salt solutions. The metals can be

Summary

Most metals react with water or steam. Potassium, sodium and calcium react vigorously with cold water to give the metal **hydroxides** and hydrogen.
Magnesium, aluminium, zinc, iron and lead react when heated with steam to give the metal **oxide** and hydrogen.
A more reactive metal will displace a less reactive one from a solution of its salt. This is called **displacement**.
Most metals react with acids to make a **salt** plus hydrogen.
Aluminium forms an unusually strong oxide layer on its surface and this protects it and makes it less reactive than the bare metal.
The experiments in this unit give you a detailed, clear order of how reactive metals are:
(most reactive) potassium, sodium, calcium, magnesium, aluminium, zinc, iron, tin, lead, copper, silver, gold, platinum (least reactive).

placed in a definite order, so that any metal in the order displaces a lower metal from one of its salts.
The order was

magnesium
zinc
iron
tin
lead
copper

Any metal will displace a lower metal from a solution of one of its salts.

Reaction with acids

Again there is a definite order of reactivity when metals react with dilute sulphuric acid. For those metals that react, hydrogen is given off and a salt is formed. Copper does not react.
The order of reactivity is

magnesium
zinc
iron
tin
lead
copper

Questions

1 **a** Describe how potassium reacts with water.
b Write a word equation for the reaction.
c What sort of substance is formed in the water?
d How is the substance formed in the water tested?
e What gas is given off?
f How is the gas tested?

2 You have decided you want to leave your own personal 'time capsule', so that people in the distant future will know what it was like to be a teenager in the twentieth century. One item in the capsule will be a metal plate, inscribed with your personal details. Which metal would you choose to make the plate from, making your choice from the list below?

magnesium
aluminium
zinc
iron
copper

Explain why you made your choice.
For each metal you did *not* choose, give a reason for not doing so.

3 You have been given a piece of greyish coloured metal which can be handled safely.
Suggest four possible metals which it might be.
You think it looks most like a piece of zinc. From the reactions you have observed in Units 6 and 7, describe how you would try to prove that it is zinc. Note any results you would expect to obtain if it is zinc.

Unit 8
How do we obtain metals?

Metals are obtained from the Earth. A few, such as silver, gold, and to a small extent, copper, are found free in nature. We say such metals occur **native**. But most metals occur as impure compounds in rocks. Rocks containing metal compounds, from which metals are obtained, are called **metal ores**.

In terms of usefulness and production tonnage, iron, aluminium and copper are world leaders. Examples of their ores are shown in Figures 8.1, 8.2 and 8.3.

Haematite is a form of iron oxide. It is called haematite because of its blood red colour. The red pigment in blood is called haemoglobin. Typically it contains between 30% and 60% of iron. The Earth's crust contains about 5% of iron. Over 600 million tonnes are produced worldwide each year.

Bauxite is a form of aluminium oxide. Typically it contains about 28% aluminium. Aluminium is the commonest metal in the Earth's crust, about 8%. About 13 million tons of aluminium are produced worldwide each year, from bauxite.

Chalcopyrite is a compound of copper and iron with sulphur. It contains only about 0.5% copper. It has a yellow glint like gold, and it is sometimes called fools gold. About 8 million tonnes of copper is produced in the world each year.

Metals are not extracted from all metal-bearing rocks. There are several reasons for this, for example:

- The rock may contain too small a percentage of the metal. A lot of rock would have to be processed to obtain a little metal, making the process uneconomical.
- The rock may be in a place which is remote and difficult to reach, such as in the Antarctic.
- The rock could be near the surface in an area of outstanding natural beauty, where excavation and the constant movement of heavy lorries would destroy that beauty.

As the Earth's reserves of metals begin to run out, however, we may have to start thinking about extracting metals from such rocks. Figure 8.4 shows the estimated lifetime of some metals.

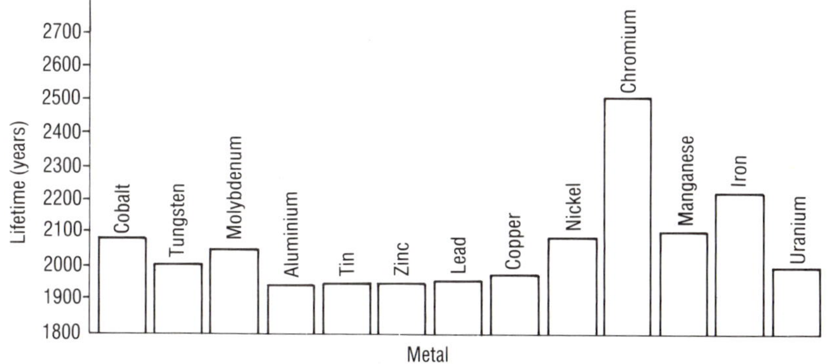

Figure 8.4 The estimated lifetime (years) of some metals

Figure 8.1 Specimen of banded ironstone (haematite-quartzite)

Figure 8.2 Bauxite – a form of aluminium oxide

Figure 8.3 Chalcopyrite ('fools gold') – a compound of copper and iron with sulphur

Questions

1 The pie-chart shows the abundance of elements in the Earth's crust.
(i) Which metal is the most abundant?
(ii) Which metal is more abundant than potassium, but less abundant than magnesium?
(iii) Which one of the three most widely-used metals accounts for less than 2% of the Earth's crust?

2 The figures below show how the production of aluminium and copper have grown since the end of World War II.

Year	Aluminium 1000 tonnes	Copper 1000 tonnes
1945	650	2150
1950	1450	2500
1955	3000	3150
1960	4700	4400
1965	6350	5100
1970	10300	6450
1975	13100	7300
1980	16050	7850
1984	13400	8250

(i) Plot graphs for these figures. Use a different colour for each metal.
(ii) What seems unusual about the production of aluminium between 1980 and 1984?
(iii) In what year did the annual production of aluminium overtake that of copper?
(iv) What was the annual production of copper in 1973?
(v) Make an estimate of the world production of copper in 1990. (If you are reading this after 1990, try to find out the actual figure, and see how close you are.)

3 Use the chart on page 37 to answer the following questions.
(i) Which two metals are expected to last the longest?
(ii) Of the three most widely-used metals, which one is expected to last the longest? Which one is expected to run out first?
(iii) How much longer are our supplies of nickel expected to last?

4 (i) There are several ores of iron, other than haematite. Find out the names of as many as you can, and where they are found.
(ii) What is bauxite? Find out why it is called bauxite.

In order to obtain a metal from its ore, the metal has to be separated from the other elements with which it is combined. Reduction usually involves the removal of oxygen from a substance. Reduction of a metal ore to the metal is the reverse of reacting a metal with other substances. Since the reactivity of metals follows a definite order, does the reverse process, reduction, follow the reverse order?

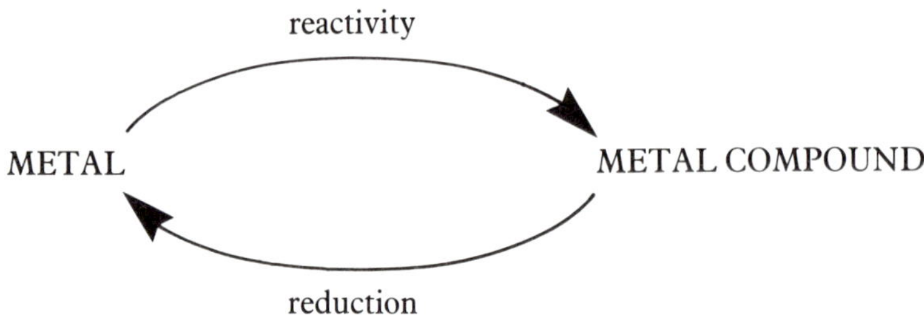

The easiest metal to extract from its ore is copper. The ore, for example chalcopyrite, is simply roasted in air, and impure copper, called blister copper, is produced.

But for other metals, extraction from the ore is more difficult. In the following units, we will look at some of the methods used.

Summary

Metals such as silver, gold and copper occur uncombined in nature (**native**).
Most metals occur in nature combined with elements such as oxygen or sulphur in rock.
Chemical reduction is used to extract many metals from their rocks (ores) – see Unit 9.
Electrolysis is used to extract the most strongly combined metals from their rocks – see Unit 10.
The Earth contains a limited amount of metals. We may run out of some metals within 10 to 20 years.

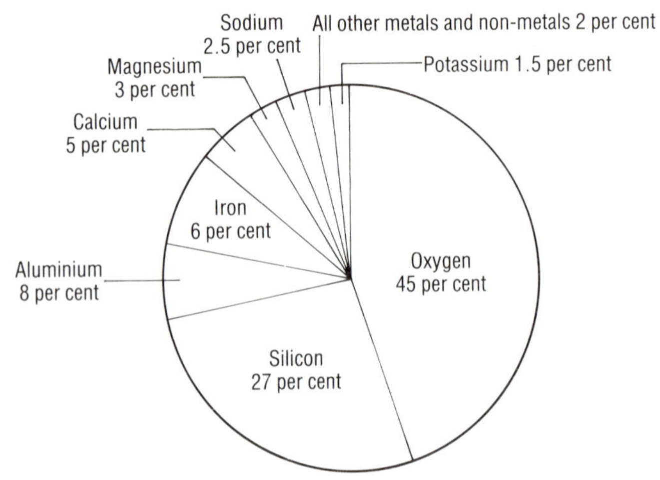

Unit 9
Reducing metal ores using carbon

Extracting iron in the blast furnace

In the blast furnace, iron ore (haematite) is **reduced** to iron. The reducing agent (the substance which reduces the iron ore) is carbon monoxide gas. This is produced in the furnace from the coke (which is mostly carbon). The carbon 'steals' the non-metal part of the ore from the metal, leaving the metal behind. The construction of the blast furnace is shown in Figure 9.1.

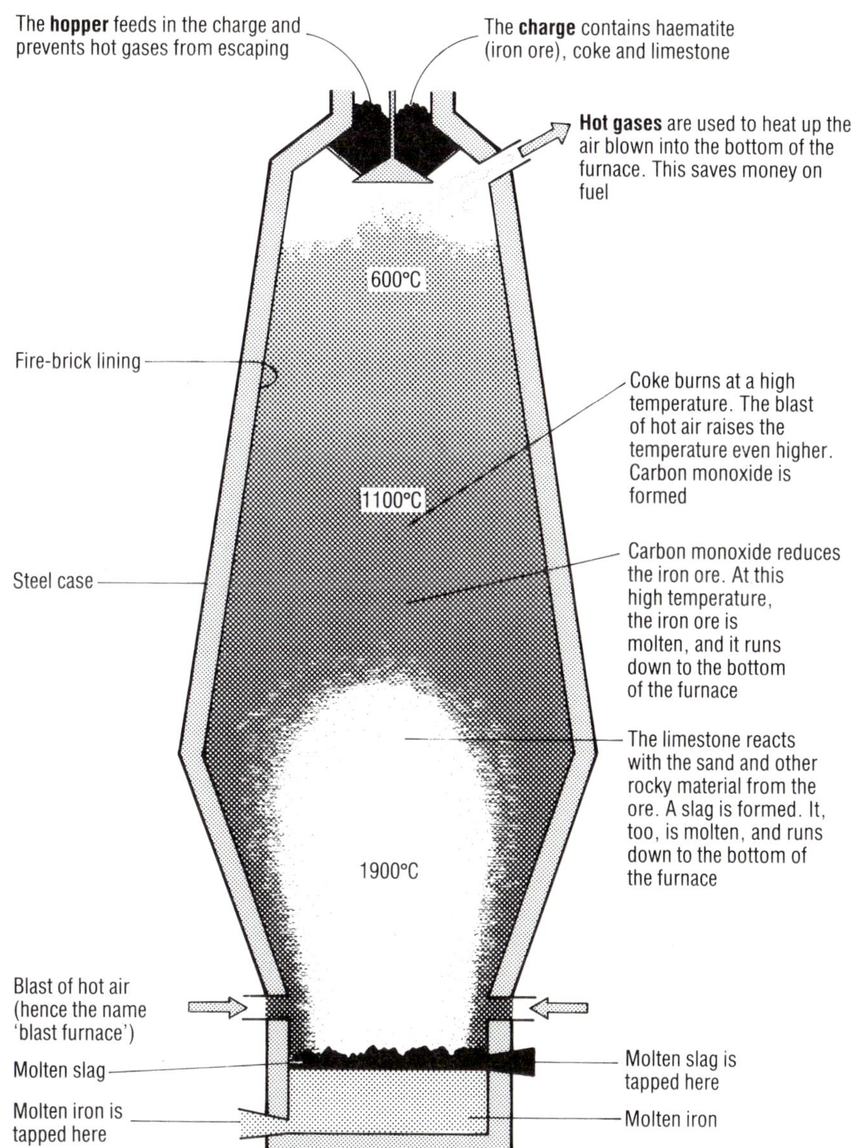

Figure 9.1 A blast furnace

Activity 9.1

To find which metal ores can be reduced by carbon

Wear safety goggles

1 Mix together some lead oxide and carbon powder.
2 Place the mixture in the depression in a charcoal block. Dampen the mixture with a drop of water.
3 Light a Bunsen burner, and open the air-hole slightly.
4 Using a blowpipe, direct a flame onto the mixture on the charcoal block. Continue for about two minutes.

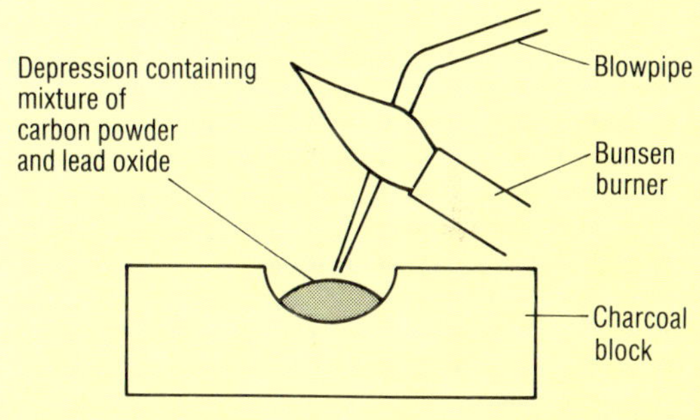

5 Look at the mixture for the shiny beads of lead.
6 Repeat the experiment with the oxides of zinc, iron and magnesium.
   - Which metal ores can be reduced by carbon?

Casting iron at the Queen Anne blast furnace, British Steel, Scunthorpe

The blast furnace is at least 30 m high. It is designed to be as economical as possible, so it runs continuously for at least 2 years. The furnace is only shut down when the firebrick lining is burned out, so that the lining can be replaced.

The chemical reactions in the furnace are:

$$\text{carbon} + \text{oxygen} \rightarrow \text{carbon dioxide}$$
$$C + O_2 \rightarrow CO_2$$

The carbon dioxide formed reacts with carbon to form carbon monoxide

$$\text{carbon dioxide} + \text{carbon} \rightarrow \text{carbon monoxide}$$
$$CO_2 + C \rightarrow 2CO$$

The carbon monoxide reduces the iron ore

$$\text{iron(III) oxide} + \text{carbon monoxide} \rightarrow \text{iron} + \text{carbon dioxide}$$
$$Fe_2O_3 + 3CO \rightarrow 2Fe + 3CO_2$$

The limestone in the charge breaks down in the intense heat

$$\text{calcium carbonate (limestone)} \rightarrow \text{calcium oxide} + \text{carbon dioxide}$$
$$CaCO_3 \rightarrow CaO + CO_2$$

The calcium oxide so formed reacts with the sandy impurities (silicon dioxide) to form calcium silicate – the slag

$$\text{calcium oxide} + \text{silicon dioxide} \rightarrow \text{calcium silicate}$$
$$CaO + SiO_2 \rightarrow CaSiO_3$$

British Steel... a profit at last

Until 1967 the iron and steel factories in Britain were privately owned. They were finding it increasingly difficult to make a profit as other countries had changed to large-scale production to keep the price of steel low. The Iron and Steel Act of 1967 meant that all but 10% of British steelmakers came under public ownership. The more profitable sections, steel-finishing and specialist steel making, were amongst the 10% left out. The 1967 Act signalled the birth of the British Steel Corporation (BSC). The idea was to shift to large-scale production near deep-water coastal locations.

In 1973 the government set aside £3000 million pounds to modernise the steel industry in five main areas, South Wales, Sheffield, Scunthorpe, Teeside and Scotland. Outdated plants were closed and the fight to compete with other countries commenced. In the mid-1970s world energy prices rose dramatically and the world demand for steel dropped sharply. BSCs plans to expand their output had unfortunately come at the wrong time. Heavy financial losses were involved which continued in the 1970s. BSC pushed for greater efficiency in the industry involving less workers. In 1980 there was a crippling three-month steel strike over pay and the pressures for change. Markets for steel were lost at a time when there was too much steel being produced in the world.

Drastic measures were required to save the steel industry. By the end of 1980 more outdated steel plants had to be closed. Locally-negotiated bonus schemes were introduced to give more money to workers if they agreed to be more efficient. The energy required to make steel was reduced. All but mainstream steel business was sold off for £705 million pounds. The price to pay for all this was a reduction in the workforce from 130 000 in 1980 to 54 000 in 1987. The drastic slimming down payed off with a profit in 1986. One problem remained… the world was over-producing steel by 30 million tonnes a year.

In December 1988 British Steel was privatised again. Such was the interest in their shares that they could have been sold three times over! The new lean British Steel was in great demand. A full circle in just 20 years.

Activity 9.2

Study the section on British Steel including the data on the steel industry below, and answer the following questions.

1 Why was the steel industry taken into public ownership?
2 What arguments can you produce for now selling off the steel industry to private enterprise? Will a sell off mean that the public ownership has been a waste of time?
3 What factors stopped the proposed growth of British Steel in the 1970s (include the beginning of 1980)?
4 List the drastic measures taken in the 1980s to revive the steel industry.
5 Plot a bar graph of the figures for profit and loss in the BSC during the 1980s.

Year	Millions of £s (− for loss, + for profit)
1981/82	−327
82/83	−383
83/84	−174
84/85	−114
85/86	+ 76
86/87	+206

6 Why were deep-water coastal locations chosen for the larger steel works?

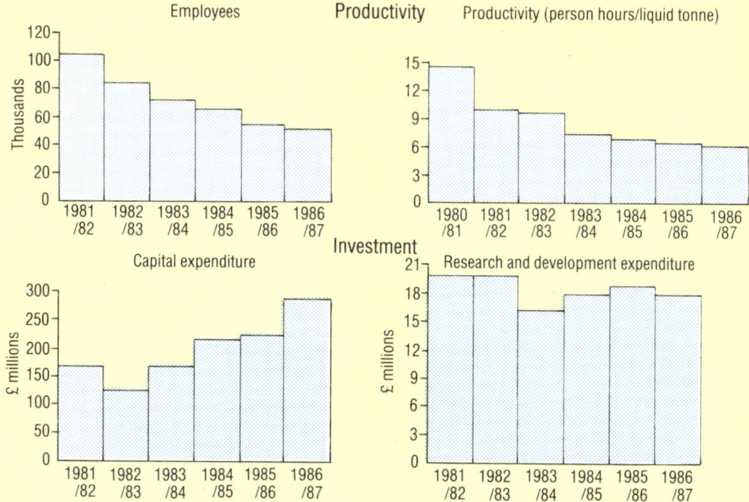

7 Look at the bar charts above. Comment on the trends in (i) the number of employees in BSC, (ii) the number of men required to make one liquid tonne of steel (productivity), (iii) the money invested in research and development and the steel industry (capital expenditure) as a whole.
8 Give your reasons for thinking British Steel is in good competitive shape. Are there any big problems on the horizon for the steel industry?

Activity 9.3

Figure 9.2 shows the relining of the number one blast furnace in Redcar. It is one of the most advanced in the world and produces 10 000 tonnes of iron a day. The table below shows the input and output for the furnace each day. Figure 9.3 shows the whole of the blast furnace at Redcar.

Figure 9.2 Relining the blast furnace at Redcar

Figure 9.3 The Redcar furnace

	Input (tonnes)		Output (tonnes)
Ore	16000	Iron	10000
Hot air	13400	Slag	3300
Oil	700	Furnace gas	22000
Coke	4500		
Limestone	400		

Answer the following questions.

1 Check the input and output totals. Give reasons why you think these totals should be the same.
2 What fraction of the ore appears as iron in the output?
3 Compare the total input with the solid output.
4 Why do you think a greater mass of gas is given off than is taken in?
5 The oil mentioned is added to the hot air blast. Why do you think oil is added?

Reduction of other ores

Zinc and lead are obtained in a similar way to iron. The ores zinc blende (containing zinc sulphide) and galena (containing lead sulphide) often occur together, so they are processed together. The ores are first roasted to convert them to the oxides. They are then mixed with coke and limestone and fed into a smaller version of the blast furnace. The carbon monoxide produced from the burning coke reduces the oxides to the metals. Molten lead and slag are tapped from the bottom of the furnace. Zinc boils at 913° C and evaporates in the furnace. It escapes from the top, and is rapidly condensed back to the solid metal.

Summary

Carbon, when heated with many ores, will take away the oxygen leaving the metal.
The blast furnace uses carbon to reduce iron oxide to iron.
The British steel industry has gone through the cycle of being privately owned, nationally owned and then privatised again, over the last 20 years.
British Steel used to lose money but since 1986 it has made a handsome profit.

Questions

1 Most of the iron from a blast furnace goes to make steel. For this reason, steelworks are usually built near to blast furnaces. Use Figure 9.4 showing the main sites of steel production by British Steel to decide whether the sites chosen have all the requirements for the economical production of steel.
   (i) Iron ore is imported in ships, so deep water is needed.
   (ii) Coke is produced from coal, so a good supply of coal is needed.
   (iii) A blast furnace and steelworks need a huge amount of energy.
   (iv) Finished products have to be taken all over the country. Is transport easy?
   (v) A fairly large work force is needed. Are there big towns near by?
   (vi) A large supply of water is needed.
   (vii) A good supply of limestone is needed.

2 Inside a blast furnace:
   (i) Non-metallic parts of the iron ore are taken away, leaving iron. What is this process called?
   (ii) What substance causes the process mentioned in (i)? Where does this substance come from?
   (iii) What is the purpose of the limestone?
   (iv) Why is it called a blast furnace?
   (v) Find out what the slag is used for.

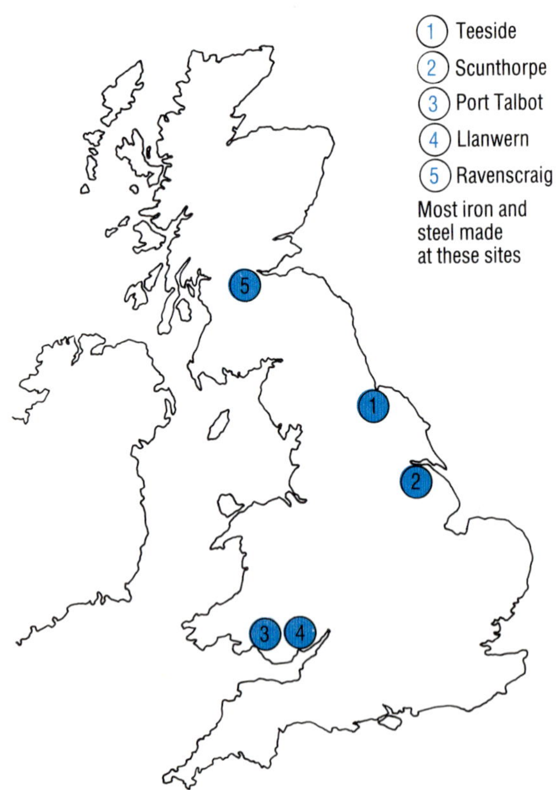

Figure 9.4 The main sites of steel production by British Steel

Unit 10
Reducing metal ores by electrolysis

Carbon isn't always the answer

Carbon is unable to take away the non-metal parts of some metal ores, including those of aluminium, magnesium, calcium and sodium. These metals hold on too tightly to their non-metal parts. A stronger method of reduction than roasting with carbon has to be used, which also makes it more expensive. It can be illustrated in the laboratory by electrolysing lead bromide to produce lead (see Activity 10.1). This method is not used industrially for the extraction of lead, as it is cheaper to use carbon.

Activity 10.1

Teacher demonstration. Reduction of lead bromide to lead by electrolysis

WARNING. This experiment should be done in a fume cupboard. Wear safety goggles.

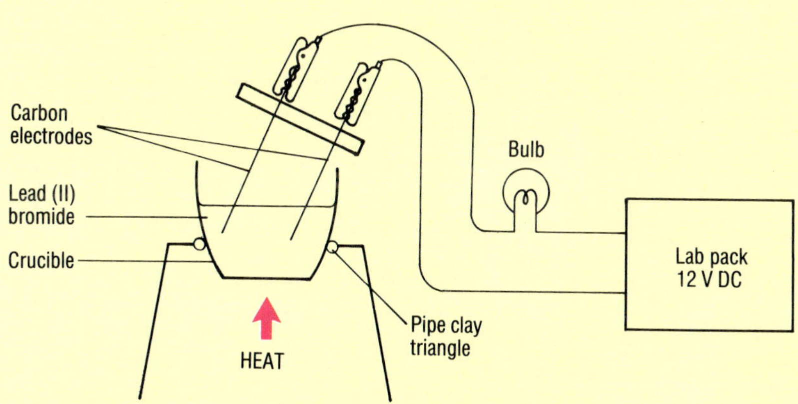

1. Arrange the apparatus as in the diagram, but do not heat immediately.
2. Switch on the lab pack. Does the bulb light? Does powdered solid lead(II) bromide conduct electricity?
3. Heat the crucible until the lead(II) bromide melts. Does the bulb light now? Does molten lead(II) bromide conduct electricity?
4. Turn off the heat and the current. When the lead(II) bromide has solidified completely turn on the current again. Does the bulb light now? Does this block of lead(II) bromide behave in the same way as the powder?

Activity 10.2

Electrolysis of copper(II) sulphate solution using copper electrodes

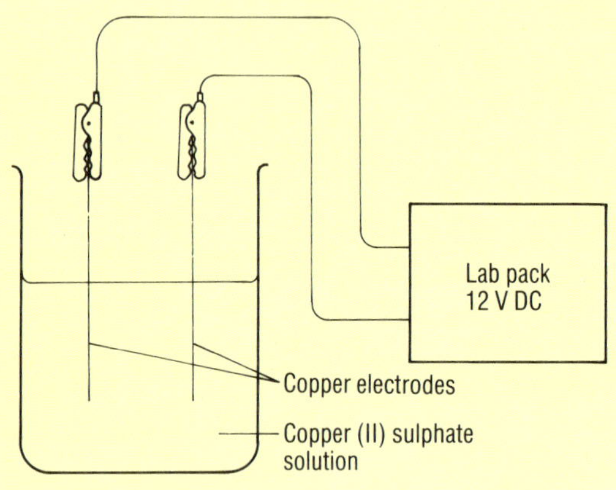

1. Weigh one of the copper electrodes.
2. Arrange the apparatus as in the diagram, with the weighed electrode attached to the positive terminal of the lab pack (it is the anode).
3. Switch on the lab pack, and leave for about two minutes.
4. Switch off and observe the cathode.
5. Remove the anode. Carefully wash it with water and then with ethanol.
   Allow the anode to dry before weighing it again.

What is electrolysis?

When an electric current is passed through a chemical and forms new substances, the process is called **electrolysis**. The chemical through which the electricity is passed is called the electrolyte. Electrolysis always requires a direct current (DC) and the presence of **ions**.

In a solid chemical, the ions are locked together in a rigid structure, unable to move. The solid substance will not conduct electricity. When the substance melts, or is dissolved in water, the ions are free to move around. The positive ions are attracted to the negative **cathode**, and the negative ions are attracted to the positive **anode**. The moving ions carry the current through the chemical.

As we have already seen, copper is fairly easy to obtain from its ore, but it contains some impurities. If very pure copper is needed, for example for use in electric circuits, it is refined (made pure) by electrolysis. (Activity 10.2 shows how this is done.)

Copper(II) sulphate is made of ions. As always, the metal ions, in this case copper, are positively charged. They are attracted to the cathode, and copper metal is deposited on this electrode.

During the process the anode loses mass. This is because copper from the anode goes into solution, forming copper ions to replace the ones deposited on the cathode.

Refining copper

The experiment in Activity 10.2 showed that copper is lost from the anode and deposited as pure copper on the cathode. This idea can be used to purify impure copper on an industrial scale. As shown in Figure 10.1, the impure copper is used as the anode. A strip of pure copper is the cathode. When the current is switched on, pure copper from the copper(II) sulphate solution is plated onto the cathode. Pure copper from the impure copper anode goes into solution in the copper(II) sulphate solution. Impurities from the anode fall to the bottom of the cell. These impurities may contain valuable metals such as gold and silver. These can be recovered from the other impurities.

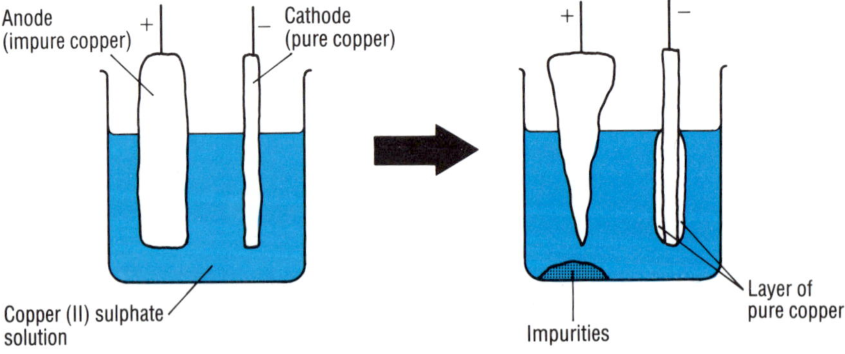

Figure 10.1 Refining copper by electrolysis

Electrolysis can plate too

In the example of the electrolysis of copper sulphate the cathode was coated with pure copper. If the cathode had been made of another metal or graphite it would still have been coated with copper. The cathode is said to be **electroplated** with copper. The metal or graphite object to be plated must always be made the cathode during electrolysis. Such a coating would protect a metal or simply make it look more attractive. Copper is not the only metal that can be electroplated onto an object. If the solution is changed to, say, chromium sulphate the object can be coated with chromium. You will have heard of chromium-plated car bumpers or cycle handlebars for example. Other metals that can be electroplated are silver, gold, platinum, nickel and cadmium.

Figure 10.2 The refining of bauxite into aluminium oxide

Industrial extraction of aluminium

Aluminium is extracted from the ore bauxite, an impure form of aluminium oxide. This is first processed to produce pure aluminium oxide. If this aluminium oxide is melted and electrolysed, it is reduced to aluminium. But the melting point of aluminium oxide is very high, 2015° C, so to melt it, and keep it molten, would require a huge amount of fuel. This makes the process too expensive. It was discovered, however, that aluminium oxide will dissolve in molten cryolite, another aluminium compound, at about 950° C. Electrolysis of this mixture still reduces the aluminium oxide to aluminium. The much lower temperature needed greatly reduces the fuel costs, making the process much cheaper. Following this discovery, aluminium became much more widely used.

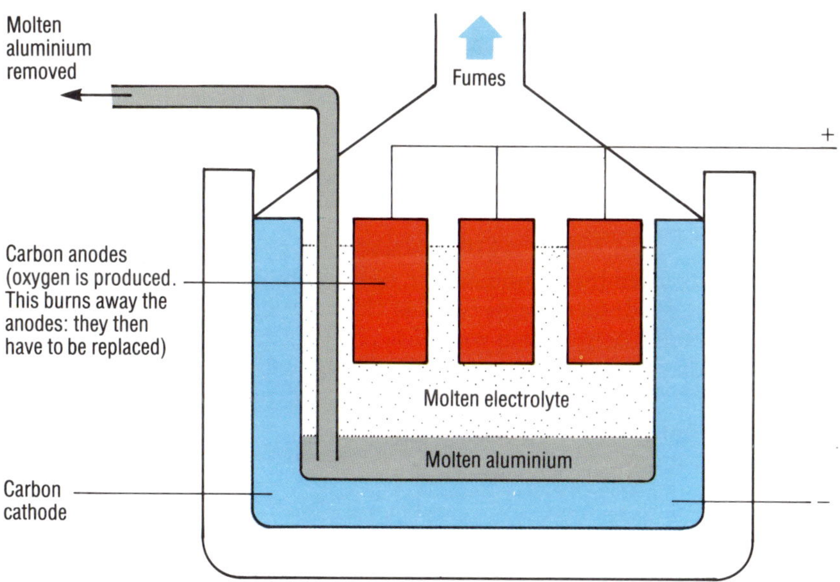

Figure 10.3 Manufacturing aluminium by electrolysis

Much of the bauxite used is mined in Australia, Guinea, Jamaica, Brazil, and Yugoslavia, but little aluminium is produced in these countries. Because of the huge amounts of electricity consumed in the process, the ore is taken to where there is a plentiful supply of cheap electricity, such as the USA and western Europe. For example, there is a big aluminium smelter (place where aluminium is extracted from its ore) on Anglesey in North Wales. It is cheaper to bring the ore to where the electricity is than to take the electrical supply to where the ore is.

Summary

Electrolysis must be used to extract strongly held metals from their ores. Potassium, sodium, calcium, magnesium and aluminium are extracted in this way. During electrolysis metals are formed at the **cathode**.
Not only extraction, but also purification of metals such as copper, silver and gold can be done by electrolysis. The impure metal is made the anode and pure metal forms on the cathode.
Objects can be **electroplated** with metals by making them the cathode during electrolysis.

Activity 10.3

The chemical input for one kilogram of aluminium

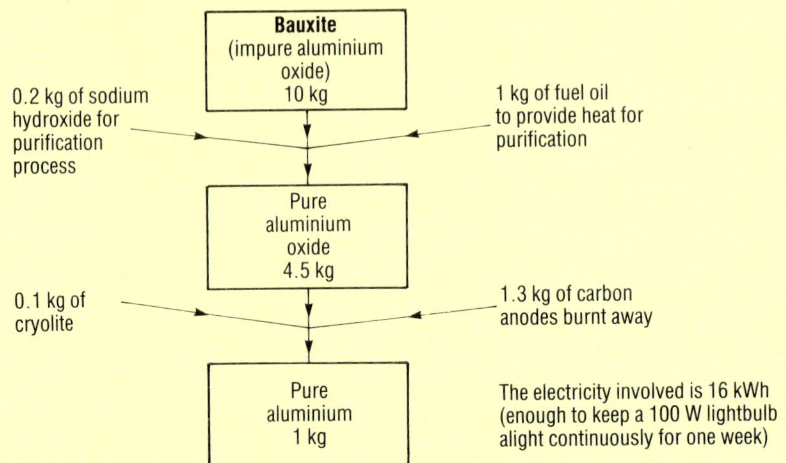

Study the diagram above and then answer the following questions.

1 What fraction of the original ore (bauxite) is converted to aluminium?
2 Calculate the mass of each chemical required to produce 10 kg of pure aluminium.
3 The aluminium oxide is separated from its oxygen (reduction) in the final stage. How is this connected with the loss of carbon anodes?
4 1 kg of bauxite is purified. Find the mass of aluminium oxide formed. How much fuel oil and sodium hydroxide is required for this stage?
5 Why is the cryolite added?

Questions

1 Sodium is extracted from common salt, sodium chloride, by electrolysis.
 (i) Will solid sodium chloride conduct electricity? Explain why, include a diagram to help explain your answer.
 (ii) What do you think is done to make the sodium chloride conduct electricity? Explain why it will now conduct. Include a diagram to help your explanation.
 (iii) At which electrode do you think sodium will be released? Explain why.

2 Explain how impure copper, obtained by roasting the ore chalcopyrite, is purified by electrolysis.

3 In the extraction of aluminium:
 (i) Which ore is aluminium extracted from?
 (ii) Aluminium oxide is produced from the ore. Why is molten aluminium oxide not electrolysed directly?
 (iii) What is done with the aluminium oxide instead?
 (iv) What are the electrodes made from?
 (v) Why is aluminium not usually extracted from its ore in the country where the ore is mined?

Unit 11
What has metal extraction to do with the reactivity series?

The diagram below summarises what we have learnt about metal extraction in the preceding units.

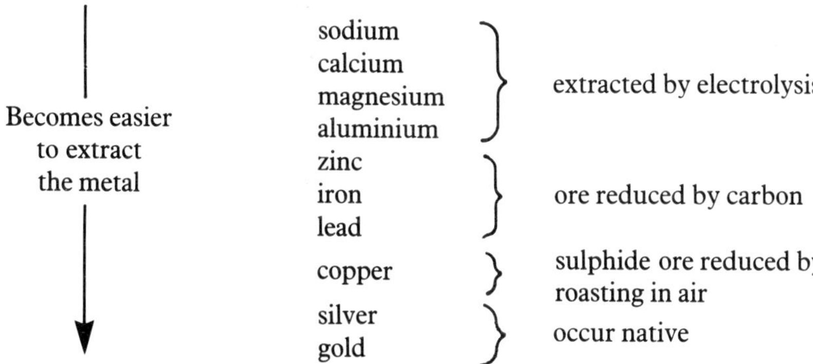

From this information, it is clear that the answer to the question posed at the end of Unit 8 is:

Yes, the ease with which metal ores can be extracted is in the reverse order to the reactivity of the metals.

Use of the Thermit process in metal extraction

In Unit 6, we saw how a more reactive metal, aluminium, can take away oxygen from a less reactive metal, iron.

This idea is used in the extraction of metals such as chromium and manganese. The ore of each metal is first converted to the oxide. It is then heated with aluminium powder, which takes away the oxygen, leaving the metal.

aluminium + chromium oxide → aluminium oxide + chromium

$$2Al + Cr_2O_3 \rightarrow Al_2O_3 + 2Cr$$

When were metals discovered?

The metals copper, silver and gold occur uncombined in nature (they are said to occur 'native') and so are most likely to have been discovered by early humans. Certainly Stone Age people must have found that these metal 'rocks' melted in their camp fire leaving nuggets of shiny metal in the cold remains of the fire. Copper, silver and gold must have been known by the year 5000 BC. They would have been used as jewellery because of their attractive colours, shiny nature and long-lasting unreactive properties. Gold was already skilfully used by the Egyptians in the year 3500 BC. Copper was found to be harder than silver and gold and became even harder when hammered for some time. Copper could be used for hunting tools instead of stone.

Around 3500 BC people would have discovered that a shiny attractive green stone could yield a red-brown nugget of copper. The green stone when heated in a fire fuelled by wood made copper by the process called smelting. The hot charcoal from the firewood reduced the copper ore to copper.

Stone Age people would, almost certainly, have discovered that some stones in the charcoal camp fire made a stronger form of copper. This stronger form of copper was called bronze and its discovery signalled the beginning of the Bronze Age and the end of the Stone Age. Bronze was really a mixture (an alloy) of two metals – copper and tin. Certain rocks contained both tin and copper and these combined in the smelting process. Bronze was much better for making lasting sharper tools for obtaining food and fashioning natural materials. Bronze Age people will have constructed primitive furnaces to extract the metals from rocks.

Lead metal would have been obtained by smelting an attractive silvery shiny ore in the charcoal camp fire. This metal must have been something of a disappointment because of its soft nature. The Romans used lead for water-piping and boilers to heat bath water.

The Iron Age started in 1200 BC when rocks containing iron were placed in a hot charcoal fire. The iron obtained from such a fire was wrought iron and this was superior to bronze for tool making. People in the Iron Age would have discovered that the properties of the iron change if the hot metal is plunged into cold water (**quenching**) or heated and slowly cooled (**annealing**). Also, combining quenching and annealing (called **tempering**) made the iron change. Iron could be softened for working or hardened for tools by such processes.

Iron Age people also discovered that if the camp fire was particularly hot or the iron was left in the fire for extra time, the iron was harder. This harder iron is called **cast iron** today.

Iron and steel became particularly useful during the Industrial Revolution, 1790 to 1860, when tools and machines were largely made from this strong hard metal.

The most reactive metals took ages to discover simply because the camp charcoal fire could not reduce their ores. Aluminium was not discovered until AD 1827. Potassium, sodium, calcium and magnesium were extracted only in recent times.

Summary

The method used to separate metals from their ores depends on how reactive the metal is.
Metals were discovered at about 5000 BC. By the year 3500 BC metals were being reduced in charcoal fires. Bronze then iron were the metals used to replace stone for tools.
The most reactive metals were the last to be extracted from their ores.

Questions

1 Find out which metals, other than chromium and manganese, are extracted using the Thermit process. For each metal, find out the name of an ore from which it is extracted, and the name of one place where the ore is found.

2 Which of the following do you think is the most likely date for the discovery of potassium?
a) 1807 AD b) 892 AD c) 500 BC d) 5000 BC
Give reasons for your answer.

Unit 12
Corrosion

What do the metal items shown in Figures 12.1, 12.2 and 12.3 all have in common?

Figure 12.1

Figure 12.2

Figure 12.3

Answer: all of them have corroded. That means they have all undergone reactions with substances in their surroundings, which has caused a change to the metal. The steel from which the car is made has rusted. This is serious, and if it is not treated early enough, very costly, since the car will have to be scrapped.

The statue is made of bronze, which contains copper. The dome on the building is made of copper. In both cases, the copper has reacted with gases in the air, and has formed a green layer of copper carbonate on the surface, which is called **verdigris**. In contrast to rust on articles made from iron and steel, verdigris is quite attractive, and protects the copper from further corrosion.

Contrast the items in the pictures above, with those in Figures 12.4 and 12.5.

Figure 12.4

Figure 12.5

Neither of the metals in Figures 12.4 and 12.5 has corroded. Why do some metals corrode, while others don't?

The reactivity series provides the clue.

Metals from the top of the reactivity series, such as sodium and calcium, are not used to manufacture items. They are far too reactive, and would corrode rapidly.

Metals like silver and gold, from the bottom of the series, are very unreactive, and do not corrode. Silver tarnishes slowly, but this is only a very slight form of surface corrosion, and it can be easily removed by polishing. Jewellery made from these metals lasts for thousands of years, and still retains its attractiveness. You wouldn't want to pay hundreds of pounds for a beautiful ring that would have corroded away within a few years!

The aluminium on the building has not corroded either, and will last as long as the building itself, although aluminium is quite high in the reactivity series. We have already seen in Unit 7 that aluminium is normally protected by a thin coating of its oxide, which prevents corrosion. Only when this layer is removed chemically will the aluminium corrode rapidly. This never happens under normal circumstances.

Rust

The costliest form of corrosion is rust, which affects iron and steel.

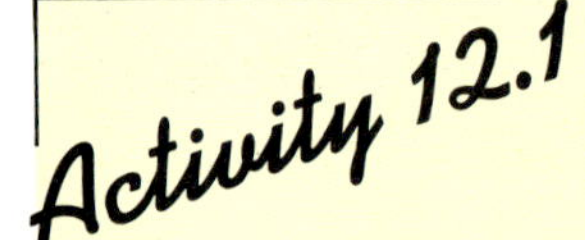

What causes rust?

Set up this experiment, and leave it for a few weeks. Inspect it frequently, and record your observations. The nails should be shiny steel.

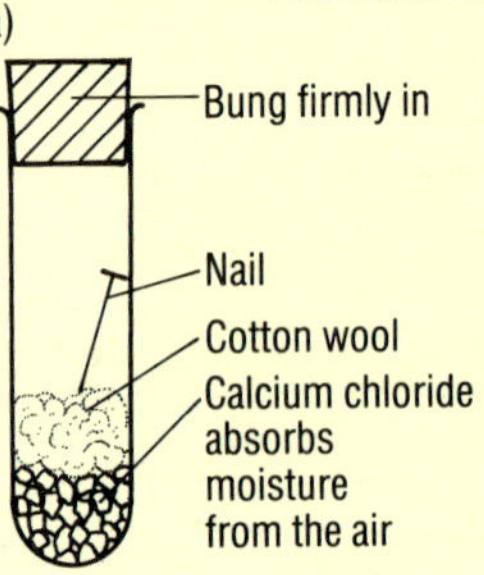

1 The nail in this test tube is in contact with dry air only.

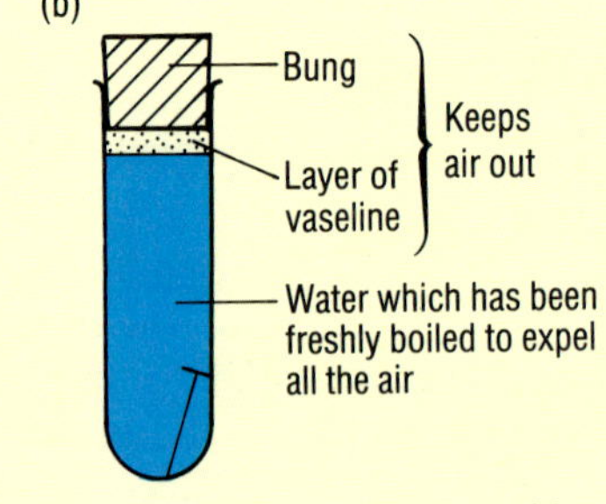

2 The nail in this test tube is in contact with water but no air.

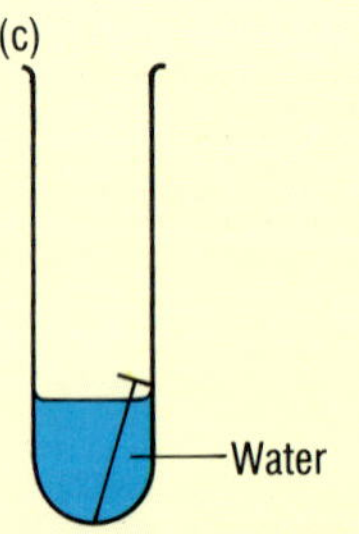

3 The nail in this test tube is in contact with air and water.

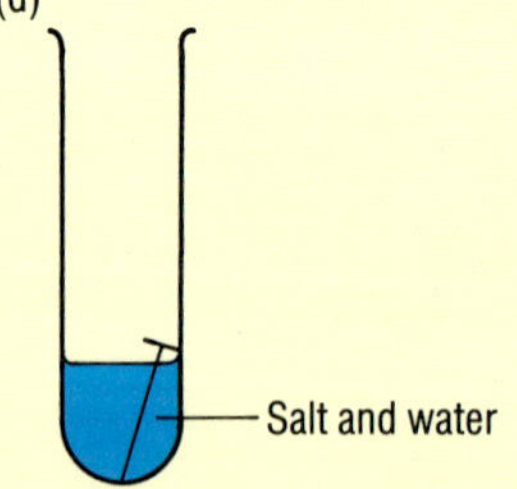

4 The nail in this test tube is in contact with salt, water and air.

Find out if dilute acid affects the rate of corrosion.
Repeat the experiment but this time storing the tubes in (i) a refrigerator and (ii) near a warm radiator.

1 Does air alone cause rusting?
2 Does water alone cause rusting?
3 What conditions are needed for rusting to occur?
4 Does salt accelerate rusting?
5 Does the temperature affect the rate of the rusting process?

Both air and water are needed for rusting to occur (see Activity 12.1). The presence of salt speeds up rusting. Rock salt is spread on the roads in winter to melt ice and snow. While this makes driving safer, it can also make it more expensive by helping to make the cars and other vehicles rust faster. Steel structures near the coast, where the air often contains sea-spray, are more at risk from rusting than similar structures inland. But steel structures which are in the sea, such as ships and oil-rigs, are under the greatest threat from rusting.

Dilute acid also increases the rate of corrosion as shown in Activity 12.1. Sulphur dioxide gas in the air forms a dilute acid when it dissolves in rain-water. The sulphur dioxide is caused by industries which burn fuels such as coal and oil which contain sulphur. The sulphur combines with oxygen from the air during burning, forming sulphur dioxide. This is released into the atmosphere, and acid rain results.

Figure 12.6 This oil rig is protected from corrosion because it is coated with zinc. Because zinc is a more reactive metal than steel, the zinc corrodes instead of the steel. This is called sacrificial protection

Preventing rust

One of the commonest ways to prevent rusting is to paint the iron or steel. Figure 12.8 shows a car being spray painted. This prevents air and water reaching the metal. Steel can also be covered with oil or grease, for example in moving machinery parts, to exclude air and water.

When iron or steel is in contact with a metal which is above it in the reactivity series, rusting is slowed down, or even stopped (see Activity 12.2). This is because the more reactive metal corrodes instead of the iron. The more reactive metal sacrifices itself to save the iron; this is called **sacrificial protection.** On the other hand, if the iron is in contact with a metal which is below it in the reactivity series, such as copper or tin, the more reactive iron corrodes even more rapidly.

Figure 12.7 Sulphur dioxide released into the atmosphere by some industries dissolves in rainwater, forming a dilute acid. This results in acid rain which causes corrosion

Figure 12.8 This car is being spray-painted to prevent the rusting of its metal parts

Activity 12.2 When iron is connected to other metals, do they affect rusting?

Set up the following experiment. Leave it for a few weeks, and observe it every day.

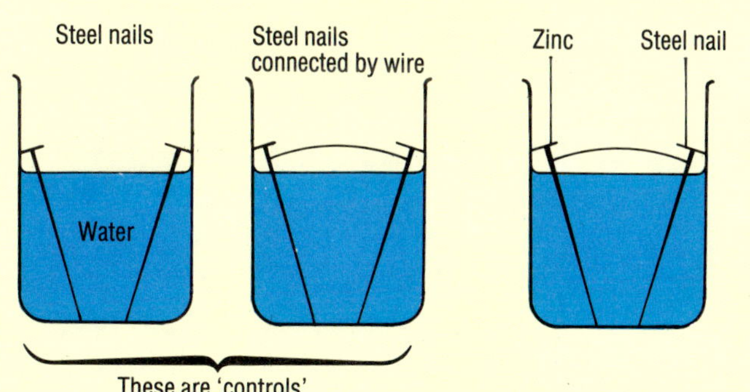

You could set up other beakers with different metals in place of the zinc and copper.

1 What is the purpose of the control experiments?
2 Which nail rusts the most?
3 Which nail rusts the least?
4 Where is zinc in the reactivity series compared to iron?
5 Where is copper in the reactivity series compared to iron?
6 What conclusion can you draw from this experiment?

Activity 12.3

Obtain a piece of a 'tin' can, and a piece of galvanised steel. Scratch the coating of each, so that the steel underneath is exposed. Leave both outside for some time, and observe them every day. Which one corrodes faster? Why?

Figure 12.9 Stainless steel is steel alloyed with 18% chromium and 8% nickel

Figure 12.10 Tin cans are really made out of steel with a thin coating of tin for protection

Galvanized into action

The largest and most advanced shopping centre in Europe was opened in October 1987. It is called the Metro Centre and is sited in Gateshead. It covers over 150 acres of once derelict land. It houses 33 shops, a 15 acre lake, hotels, a ten-screen cinema and a fantasy funland. Such a development demands a huge car park. The two multi-storey car parks cover five levels and will take over 3000 cars.

The car parks are made from 1000 tonnes of hot-dipped galvanized steel. The galvanized structural steelwork will require little maintenance during the next fifteen years. Galvanizing steel, by dipping it in molten zinc, certainly prevents corrosion.

Summary

Metals usually combine with oxygen in the air to form metal oxide corrosion.
The least reactive metals corrode less easily.
Rust is hydrated iron(III) oxide and requires iron, oxygen and water for it to form.
Rust is particularly damaging because once formed it easily falls off the metal leaving iron to oxidise further. Rust can be prevented by paint, grease (oil), electroplating, galvanising or attaching a 'sacrificial' metal.
Galvanised iron is made by coating iron with zinc.

Questions

1 Find out where platinum is in the reactivity series. Would you expect it to corrode easily? Why?

2 In the experiment shown in the diagram, in which beaker would you expect the nail to corrode the fastest? Why?

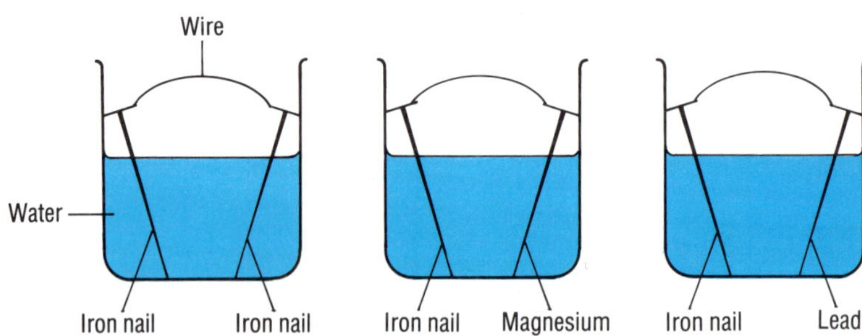

3 Find out how the following are protected from rusting
  a The Firth of Forth railway bridge.
  b The underside of a car.
  c The submerged part of an oil rig.

4 Zinc provides better protection of steel from corrosion than tin. Find out why tin is used instead of zinc to protect food cans.

5 Which of the following metals would you expect to provide sacrificial protection for steel? Why?
  (i) Magnesium.
  (ii) Lead.
  (iii) Copper.
  (iv) Aluminium.

Unit 13
Metals and portable power

Supplying power

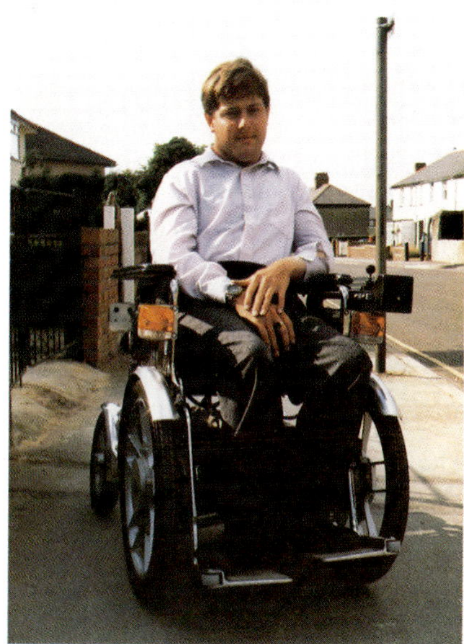

Figure 13.1 All the machines in these pictures are battery operated

All the machines in Figure 13.1 run off an electrical supply that they carry around with them. The needs of each machine are different.

The milk float and fork-lift truck need large amounts of power for relatively short periods of time – usually just a few hours each day. When the machines are not in use, their electrical sources – their batteries – can be re-charged.

The watch needs a tiny amount of electricity that must remain constant for a year or more, and its battery must be small enough to fit inside the watch.

The 'ghetto blaster' needs more power than the watch, but a lot less than the milk float. Its power supply does not need to be as small as that of the watch. To satisfy these and many other requirements, a wide range of different batteries has been developed, some of which are shown in Figure 13.2.

Figure 13.2 A selection of batteries currently available

Activity 13.1

Metals helping to produce electricity

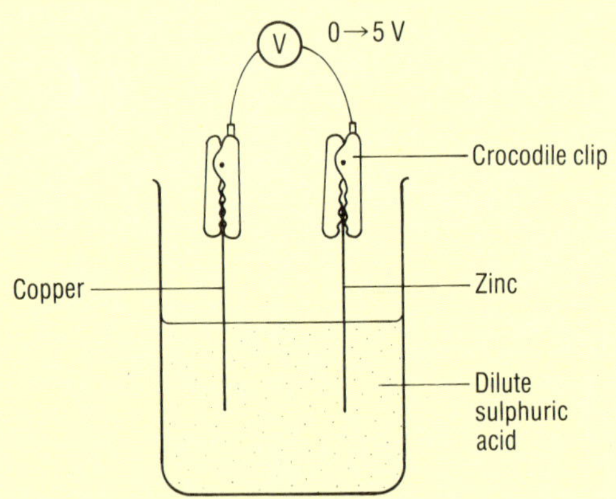

1. Arrange the apparatus as shown in the diagram. Note the reading on the voltmeter. If it gives a negative reading, connect the metals to the opposite voltmeter terminals.
2. Remove the zinc from the acid with tongs, rinse it in water, then place it on a paper towel and dry it.
3. Repeat the experiment, using other metals in place of the zinc, such as lead, tin, magnesium and aluminium.
4. Place the metals in order, the one giving the highest voltage first. How does your list compare with the reactivity series?
5. What happens if you use two pieces of copper?

The process of development continues as ever-more efficient ways of providing portable power are sought. For example, a battery-driven car would be cleaner, quieter and easier to maintain than a petrol or diesel car. But existing batteries are heavy and give the car only a limited range before they need re-charging. Scientists are continuing to search for ways of producing smaller, lighter batteries that will give cars a greater range than batteries which are currently available. With a petrol or diesel car, a tankful of fuel may take you, say, 300 miles, before it runs out. When this happens, you simply drive into a filling station, fill up, and you are ready to drive on after a few minutes. But a battery-driven car wouid have to stand motionless for several hours waiting for its batteries to be re-charged once they have run flat.

Although there are many different batteries designed to do a wide range of jobs, one thing they have in common is – metals.

Cells and batteries

What we normally call a battery should, in most cases, be called a cell. A 'battery' consists of two or more cells joined together. The 1.5 V 'batteries' you find in shops should really be called 'cells'. The 3 V and 6 V batteries are true batteries; the 3 V battery contains 2 cells and the 6 V battery four cells.

Inside a simple cell

The basic design of most 1.5 V cells (or batteries) is shown in Figure 13.3. Two pieces of different metals (or one metal and carbon) are in contact with a fluid chemical. For use in portable radios, torches, etc, it was necessary to develop a battery which did not contain a liquid like sulphuric acid, which might spill and cause burns. Instead a special paste was developed for use in the so-called dry cell, which is still the most widely used 'battery'. A single such unit should strictly be called a cell. A battery is two or more cells joined together, as in the car battery (see Figure 13.4).

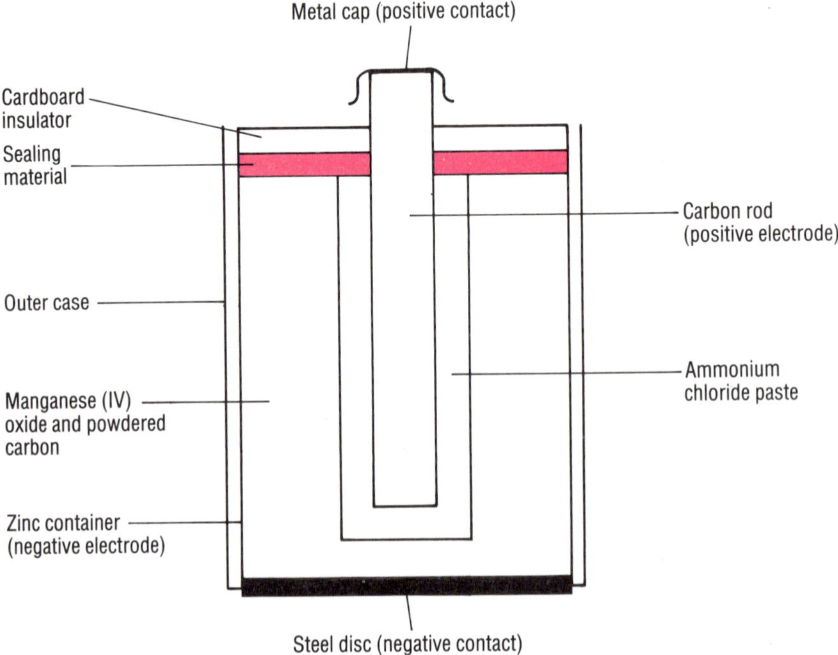

Figure 13.3 A dry cell

Lead-acid batteries

Diagram **1** of Activity 13.2 shows a simple lead-acid cell being 'charged'. One of the electrodes becomes coated with lead oxide. (Which one? What happens at the other electrode?) This now acts like a simple 2 V cell. As it is used (ie as it discharges) the plates become coated in lead sulphate. However they can be restored again by recharging. You could repeat this process over and over again – the cell is rechargeable. It is sometimes called an accumulator, because it accumulates (collects and keeps) a charge. A single lead-acid cell produces 2 V. A 12 V car battery contains six cells.

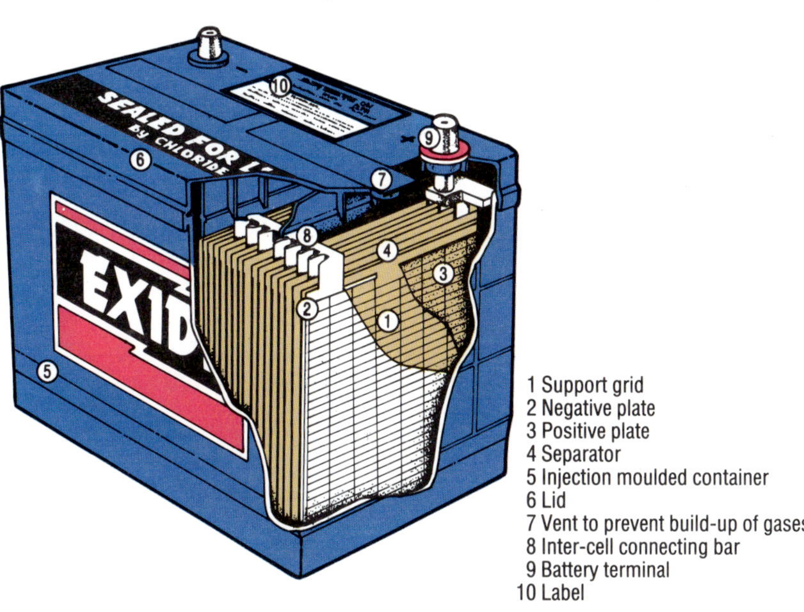

Figure 13.4 The parts of a car battery – two of the six cells can be seen

Activity 13.2

To make a lead-acid cell

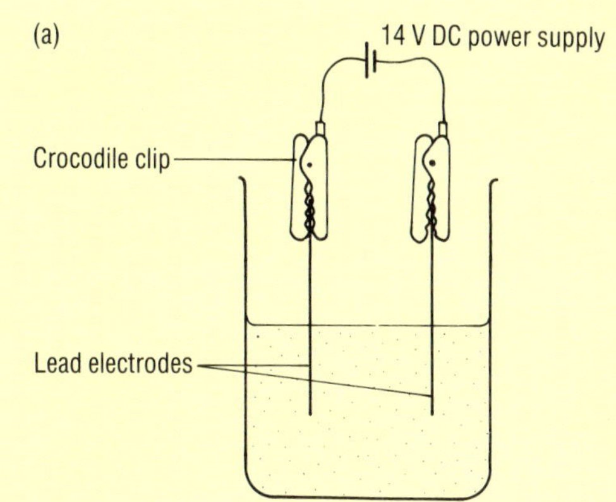

1 Arrange the apparatus as in the diagram. Switch on for one minute. Avoid inhaling the fumes.

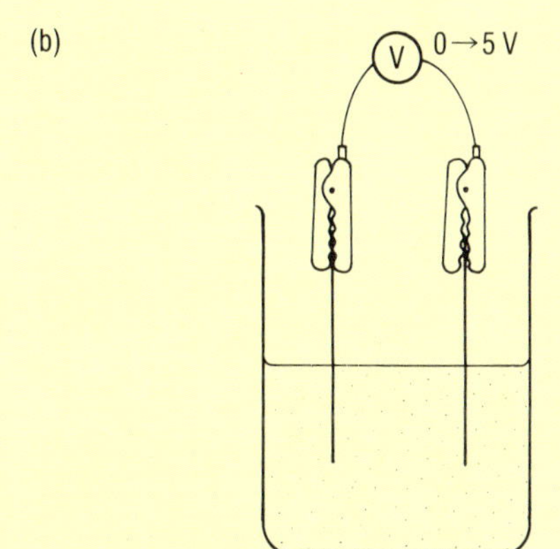

2 Disconnect the power supply. Replace it with a voltmeter. What reading do you get? You could connect a small bulb into the circuit instead of the voltmeter.

Batteries for vehicles

The battery industry has been the major consumer of lead and lead alloys for many years. The demand for lead-acid batteries has increased with the steady rise in the number of vehicles. Designers have been trying to reduce the size and weight of batteries with some success. This has meant the use of less lead in the battery while keeping good electrical performance. Smaller and lighter batteries mean more scope for the car designer.

Changes in the designs of the lead grids and the separators have meant that the battery can have less lead and a lower electrical resistance. We need a good battery to start the car on a cold and frosty morning. The engine starting power of a battery depends on the number of ampères of electrical current it can muster when the ignition key is turned; it is called the Cold Cranking Amps (CCA). The time in minutes for which a battery will discharge a current of 25 amps is called the Reserve Capacity (RC) of the battery. Some car makers are willing to allow the RC to be lower while the CCA must be maintained or increased.

The Torquestar® battery was introduced by the Chloride company in 1981. These are sealed unspillable batteries which combine any gases formed when the battery is overcharged. The battery plates (grids) are cast in an alloy of lead including calcium and tin. It is still rather heavy but it has a low electrical resistance, giving a good CCA.

The Dunlop Pulsar® battery has extremely thin grids produced by expanding (done by trapping air bubbles in the metal) an alloy strip of lead, calcium and tin. These grids are located and retained in a plastic frame. The active life of batteries has shortened with the design improvements; 42 months was the average active life in 1979 and it has reduced to about 35 months today.

Activity 13.3

Study the data on batteries for vehicles and answer the questions below.

Type of battery	**Battery weight** kg	**Battery volume** dm^3 (or litres)	**Lead weight** kg	**RC** minutes	**CCA** amps
Standard design (USA)	17.3	9.08	9.34	102	420
Lightweight design (USA)	16.2	7.75	9.66	104	660
Cathanode® (USA)	14.9	7.83	9.45	100	625
Standard design (UK)	12.0	6.06	7.15	65	360
Torquestar® (UK)	11.0	6.06	6.9	55	420
Pulsar® (Australia)	9.8	5.17	5.7	66	320

1. Compare the USA and UK standard batteries pointing out the good things about each. Think about American cars and decide why their standard battery is like it is.
2. Imagine that you own a very light small car. Explain why you think the Pulsar® battery would be best for the car.
3. Imagine that you own a heavy truck that has been very difficult to start on cold mornings. Which battery would you change to from the list? Give your reasons for choosing the battery and any worries you have about it.
4. Using the data in the table compare the Cathanode® battery with the Lightweight (USA) battery. Is there much to choose between them?
5. How do you think future batteries will vary from those shown in the table?
6. A disabled person has the latest lightweight wheelchair. Choose the best battery for this vehicle and give reasons for your choice.

Figure 13.5 Car batteries

Figure 13.6 The latest in battery-powered wheelchair technology

New alloy helps mobility

A new zinc alloy, ZA27, has helped to make a battery-powered wheelchair that does more. It will climb up most hills, managing slopes of up to 30%. It can mount obstacles 11 cm high, successfully coping with most kerbstones. It can travel over rough ground and make good progress on the flat.

The ZA27 alloy contains 27% aluminium, 2% copper and 71% zinc with just a trace (0.01%) of magnesium. The castings of this alloy are used in the bracket fixing the seat to the powered wheel unit (a most important part), the wheel rims, the rear wheel forks and the control box housing. The finish of the alloy looks good too.

Activity 13.4

What properties would you look for in a metal for a battery-powered wheelchair? Make a list of your suggestions and say why you have chosen them.

The table below shows the properties of ZA27 and some other common metals and alloys.

Property	ZA27	Aluminium	Brass	Bronze	Cast iron
Density (g cm^{-3})	5	2.7	8.5	8.9	7.3
Melting range (° C)	375-484	515-605	925-940	855-975	above 1200
Tensile strength (MPa)	440	262	230	240	214
Hardness (Brinell)	30	20	12	15	50
Impact strength (Joule)	60	5	15	11	70

If you had the choice of these five materials to make a modern wheelchair, debate why, on balance, you would choose ZA27 (use the data to support your arguments).

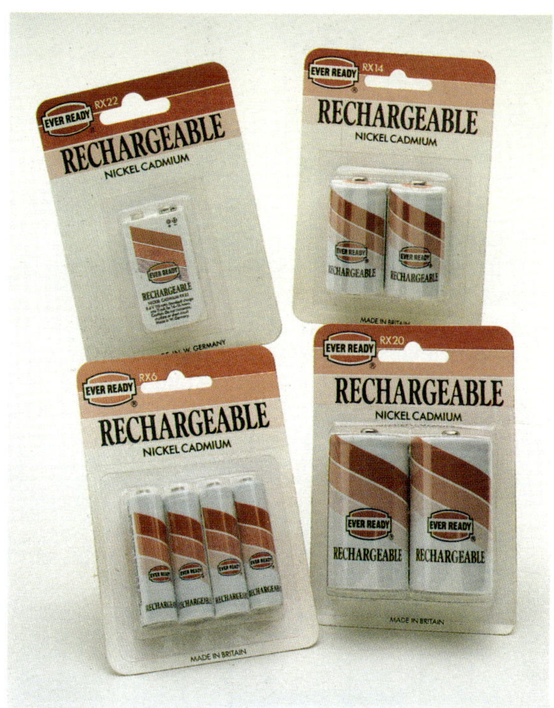

Figure 13.7 Nickel-cadmium rechargeable batteries

Other types of batteries

Although the zinc-carbon cell is still the most widely-used 'battery', several different types of cell have been developed in recent years using different metals such as nickel and cadmium. Some of these are **rechargeable**. It is important that only those cells marked as rechargeable are recharged. Trying to recharge non-rechargeable batteries is highly dangerous and could cause an explosion.

Save with rechargeable batteries

We are all familiar with the conventional battery for toys, torches and such like. They have to be thrown away after what often seems a short life. What a waste of chemicals and all the energy used to make the battery!

A rechargeable battery for the uses mentioned would have to be sealed; the chemicals could not be allowed to corrode the toy, torch or radio. An unsealed battery would lose liquid (electrolyte) and need regular 'topping-up' like many car batteries. Gases, such as oxygen, given off when batteries are charged would cause a sealed battery to explode. These problems were overcome in the early 1950s when the rechargeable nickel/cadmium alkaline cell was developed. The new design recycled any oxygen given off during recharging of the battery. A safety vent was incorporated so that IF oxygen pressure did build up due to misuse, it would leave the battery safely. Such rechargeables are sealed and do not leak. Both electrodes in the battery are made of nickel-plated iron mesh. The small holes in the mesh contain the active chemicals. These active chemicals are nickel hydroxide in the positive electrode and cadmium hydroxide in the negative electrode. The liquid (electrolyte) is potassium hydroxide solution. Figure 13.8 shows the construction of the nickel-cadmium battery.

When the battery is charged, the positive electrode becomes a basic form of nickel oxide and the negative electrode becomes cadmium metal. A

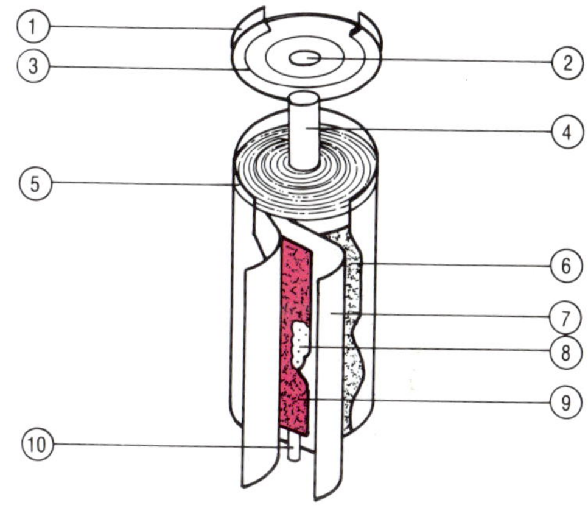

1 Nylon sealing gasket
2 Resealing safety vent
3 Nickel plated steel top plate (positive)
4 Positive connector
5 Nickel plated steel can (negative)
6 Sintered positive electrode
7 Separator
8 Support strip
9 Negative electrode
10 Negative connector

Figure 13.8 Sealed cylindrical nickel-cadmium cell

Activity 13.5

Imagine that you are to head a campaign to convince the public that rechargeable batteries are a good thing.
You have been told that sales of these batteries have been disappointing. Form a group and discuss why sales have been disappointing. (What is unattractive about these batteries?)
Outline your plan to convince the public of your case.
Design some posters to help 'sell' the idea.

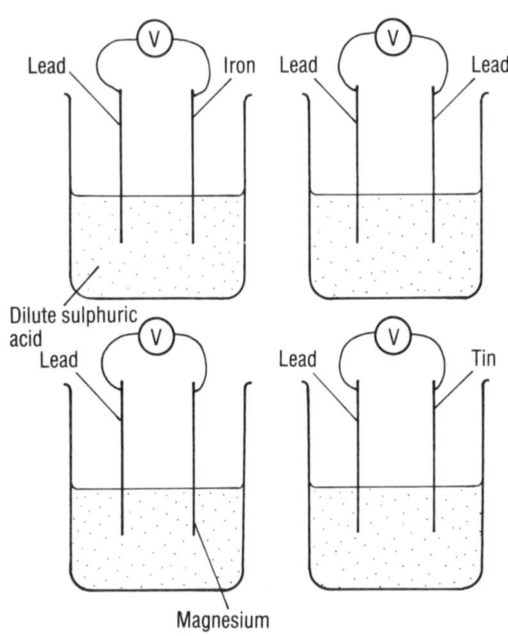

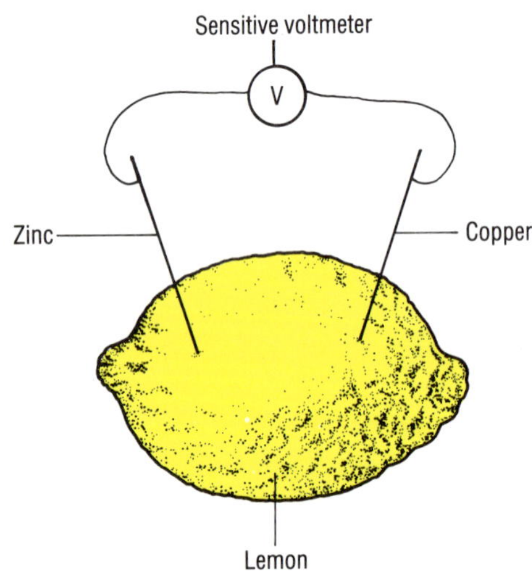

rechargeable battery will cost more initially but it can be recharged up to 500 times. Chargers can be purchased which recharge each of the four main sizes of battery. The recharging process is best when slow, and takes about 15 hours. This can work out as little as 1p only for a recharge! Rechargeables, unlike conventional batteries, keep their high voltage throughout their lives. They are generally more powerful and can be ruined and cause burns if the positive and negative electrodes are joined (short-circuiting). Adults, not young children, should be the ones to put rechargeable batteries into toys because of this extra power.

Summary

Simple **cells** contain two different metals dipped in an acid.
A **battery** is a number of cells joined together.
The dry cell contains ammonium chloride paste instead of an acid solution. It is portable and does not leak.
The car battery is a lead-acid cell and can be **recharged**.
Car batteries should ideally be light, work well even when cold, and deliver a good electrical current.
Cars/wheelchairs made of light, strong alloys can be powered more easily by batteries.
Rechargeable batteries save money and the Earth's resources. They can be used up to 500 times before being scrapped.

Questions

1 Four cells were arranged as shown. In each case, the voltmeter was connected the same way round, and gave either a positive or zero reading.
  a Which cell would you expect to give the highest reading?
  b Of the cells which give a positive reading, which one will give the lowest reading?
  c Would you expect any of the cells to give a reading of zero? If so, which one? Give reasons for your answers.
  d Give one advantage and two disadvantages of using a cell like no. 3 to produce electricity.

2 Find as many pieces of equipment as you can at home which will run on batteries. For each one, write down:
  (i) What the piece of equipment is.
  (ii) How many batteries are needed to operate it.
  (iii) What type of batteries it uses.
  (iv) What the voltage of each battery is.
  (v) What the total battery voltage is if it uses more than one battery.

3 The diagram shows an arrangement for obtaining electricity from a lemon. It works because lemons contain a weak acid. Devise an experiment to test different fruits and metals, to find which combination gives the best voltage.
Develop a plan for your experiment. It should contain references to variables such as distance between the metals, size of fruit, freshness of fruit etc, as well as the types of metal and fruit used. Decide how best to present your plan – written out normally, as a series of diagrams, as a flowchart? How will you record your results?
When your plan is ready, show it to your teacher. If there is time, you may be able to try it out.

Unit 14
Do metals have a future?

Do we need metals?

Think of your lifestyle now and the part played by metals in it. At present we all like to live long and healthy lives and presumably this will always be so. We like to be comfortable, to have some privacy and to communicate with people at other times. We want to enjoy our leisure, travel and create and see beautiful things. Metals play an important part in these activities.

The challenge of plastics

Some say that new materials are replacing metals. At present, plastics of a given strength are more expensive than metals of the same strength. Certainly plastics have replaced many metal parts in cars and the building industry. Metals are still used for the parts that require great strength, a high density and good electrical or thermal conductivity. Many new applications for metals are being found in this changing world.

Experts have calculated that by the year 2000 AD the world production of aluminium will have trebled and that of steel doubled (compared with 1980 figures). These hardly suggest that metals are being replaced!

Growth areas for metals

One huge growth area will be developing countries requiring metals for agricultural machinery, building materials, vehicles and domestic appliances. In the prosperous world there will be more leisure time for holidays and hobbies which will involve metals. Sadly, we must assume that military conflict will continue in the world and that metals will be used to make ever more deadly weapons. Future batteries will contain metals and play their part in battery-powered vehicles. Many suggest that battery-powered vehicles will be the transport of the future. We will become much more energy conscious; metals will play their part in cutting down on the waste of energy.

New and growing markets for copper

Copper is used in electrical printed circuits and this use will surely grow. In the field of energy conservation copper will play a vital part in heat-exchangers and solar heating panels.

Solar heating

An all-copper solar heating system installed in a house at Forestdale, Croydon has won a design competition. (The circuit design for the system is shown in Figure 14.1.) The solar panels are hardly noticeable as they are part of the roofing rather than placed on top of it. The circuit between the copper solar panels and the heating coil is closed so that anti-freeze can be added to the water for all-year use. The solar panels are made from strip copper and coated with black to best absorb the sun's rays. The system needs little attention as there are no moving parts. The total cost of installation was £700.

Activity 14.1

Try to project your mind into the year 2010 AD and write an essay on the role of metals at that time.

Activity 14.2

Plastics v metals in the motor car

The following table shows the growth in the use of plastics in the average car.

Year	Mass of plastic in the car (kg)
1960	10
1966	17
1972	48
1979	60
1980	85
1985	100

1. Plot a bar-graph to show this growth in use of plastics over the years. Use the graph to estimate the mass of plastic used in the year 1995.
2. How many plastic parts of a car can you think of? Make a list. The average car contains a thousand plastic parts. Which of the parts you have listed might once have been made of metal?
3. Do you think the number of plastic parts in a car will continue to increase (give your reasons)? Is the all-plastic car a possibility?
4. What are the advantages of plastic parts in a car?

Even in Croydon this will be paid for by energy savings in about ten years. In sunnier parts of the world the system would make huge savings on fuel.

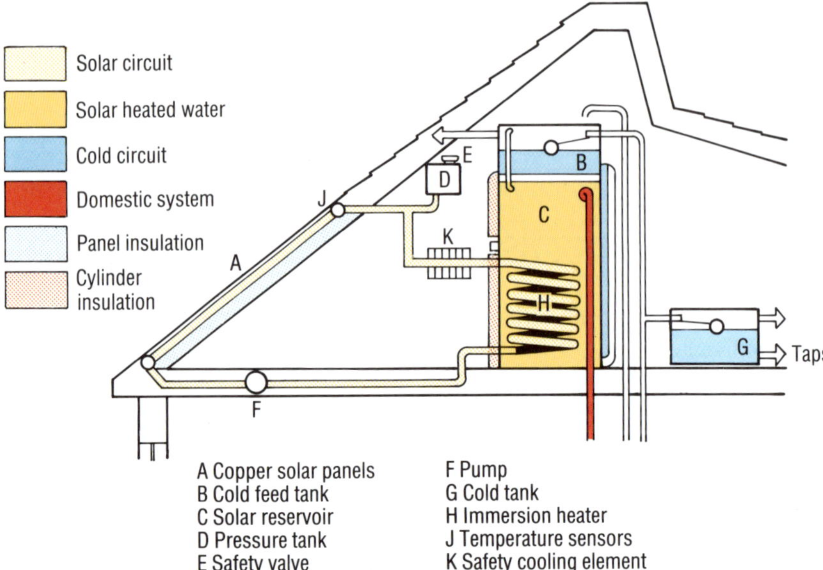

Figure 14.1 The circuit diagram for the solar heating system used in Croydon, Surrey

Figure 14.2 Domestic heat shuttle

The heat shuttle

Much of the heat made by a home or industrial boiler is lost as it travels to the air via the flue. The heat shuttle can trap and recycle this otherwise wasted energy. As shown in Figure 14.2, it consists of a bank of copper heat pipes enclosed within a casing inside the flue pipe. One end of the heat pipes rests in the hot flue gases while the other end is tapped into a water jacket. Copper is such a good conductor of heat that much of the flue-gas heat travels to warm up the water. Savings of up to 10% on fuel can be achieved using this simple device.

Copper foils algae

The concrete weir at Bridgham, Norfolk has been flame-sprayed with copper to prevent algae fouling it. Many weirs gauge the flow of water in a river and algae growth seriously affects this measurement. Algae grows prolifically on concrete in the summer months and is difficult to remove; often removal must take place every two weeks and the scraping involved can damage the concrete. Flame spraying the concrete with copper solves the problem ... no more algae sticks. The flame spraying involves copper wire being propelled into a flame or electric arc where it is atomised. Compressed air shoots the copper atoms onto the concrete surface. Copper coatings can be applied in this way to iron, steel, plastic, wood or plaster. Perhaps this will be useful in the future.

High melting alloys

It has been suggested that as the supply of certain chemicals becomes more scarce in the world many extractions are likely to involve ores which need high temperatures to separate them. Energy in the future may come from fusion. This is a process which involves temperatures of millions of degrees centigrade. The sun produces its energy by fusion.

Metals with very high melting points such as molybdenum (melting point 2622°C), niobium (2468°C), tungsten (3370°C), and tantulum (2996°C) will be required to make the alloys for the containers needed for these processes using these very high temperatures. For the highest temperatures metals will be coated with a thin layer of ceramic material to resist the heat. As a comparison the melting point of iron is only 1535° C.

Will we run out of metals?

The graph on page 37 shows the calculated lifetime of some common metals. Experts cannot be sure how accurate these figures are. Will new sources of the metal be found in Alaska? Will new uses mean that some metals will be used up sooner? To play safe we should recycle metals rather than throw them onto the scrap heap. If metals are recycled less energy is required than making new supplies of the metal.

Cans used for drinks are often made of aluminium. Aluminium is well worth recycling because huge amounts of energy are involved in making it from its ore. It takes 95% less energy to remelt used aluminium cans. No wonder the Aluminium Recycling Can-paign (as it is called) uses the slogan *'Don't trash it … cash it'!* Think of ways that you could make money and save the world's metals and energy. If you decide to save aluminium cans follow the rules shown by the 'can-paign'.

1 Test it
The magnet test. This is very important, because not all drinks cans are made from aluminium. Test the side – not the top – of each can with a magnet. Aluminium is not magnetic so keep only those cans that the magnet does *not* stick to.
Some cans have the (Al) or (Alu) symbol, those are aluminium too.

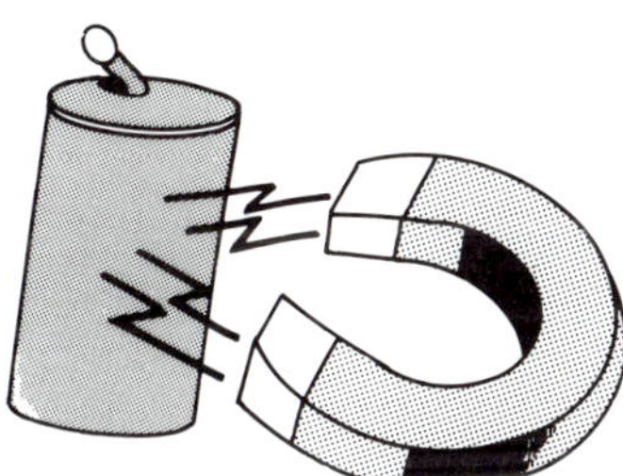

2 Crush it
To save space put the ring pull inside the can and then crush the can, before you put it in the bag.

3 Bag it
Plastic refuse bags are ideal for keeping your can collection in. Then check with your organiser what to do next. Your organiser may arrange to take away your collection for you, or ask you to take your collection to a special place.

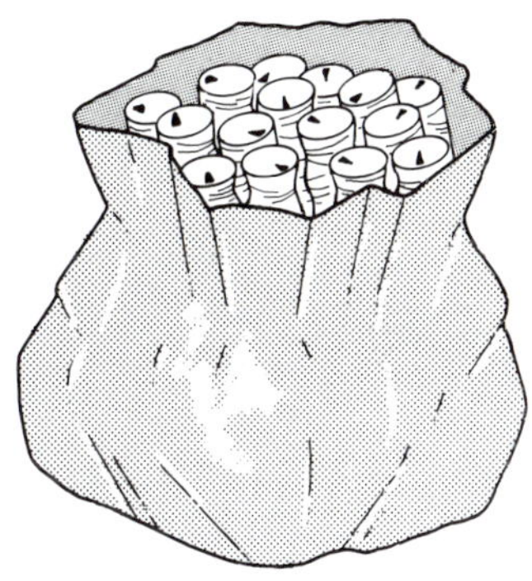

Figure 14.3 The Aluminium Recycling Can-paign

Summary

Plastics have replaced metals for many tasks.
Metals still have their unique properties which cannot be matched by other materials.
New uses are being found for metals.
Recycling saves: time
energy
natural resources
money
the problem of waste disposal
Recycling of metals is the only sane policy to prevent some metals running out in the near future.

Solutions

Unit 3
Wordsearch

S	S	M	U	N	I	T	A	L	P	N
M	T	S	Z	I	N	C	T	E	N	S
U	C	E	A	T	G	O	L	D	I	Y
I	H	S	E	R	S	P	E	L	L	R
N	R	O	D	L	B	P	V	L	I	U
I	O	D	A	E	I	E	H	E	T	C
M	M	I	Y	A	R	R	T	K	H	R
U	I	U	G	D	O	Z	O	C	I	E
L	U	M	A	G	N	E	S	I	U	M
A	M	U	I	C	L	A	C	N	M	S

Unit 5
Crossword

1 E	L	E	2 C	T	R	O	3 N		4 M	A	5 N
L			U				O		A		U
E		6 P	R	O	T	O	N		T		C
M			R				7 M	E	T	A	L
8 E	R	N	E	S	9 T		E		E		E
N			N		H		T		R		U
T			T		R		A				S
					10 E	E	L			11 H	
12 S	13 I	M	P	L	E			14 O	V	E	N
	O							N		A	
	15 N	E	G	A	T	I	V	E		T	

Index